Principles of Biology | Ecology and Evolution

Biology Lab Manual

Second Edition

Jennifer Schramm

Chemeketa Press | Salem, Oregon

Principles of Biology | Ecology and Evolution: Biology Lab Manual

ISBN-13: 978-1-955499-54-5

Chemeketa Press
Chemeketa Community College
4000 Lancaster Dr NE
Salem, Oregon 97305
collegepress@chemeketa.edu
chemeketapress.org

Cover design by Ronald Cox
Interior design by Ronald Cox and Abbey Gaterud

A full list of credits appears on page 126 and constitutes an extension of this copyright page.

Printed in the United States of America.

Land Acknowledgment

Chemeketa Press is located on the land of the Kalapuya, who today are represented by the Confederated Tribes of the Grand Ronde and the Confederated Tribes of the Siletz Indians, whose relationship with this land continues to this day. We offer gratitude for the land itself, for those who have stewarded it for generations, and for the opportunity to study, learn, work, and be in community on this land. We acknowledge that our College's history, like many others, is fundamentally tied to the first colonial developments in the Willamette Valley in Oregon. Finally, we respectfully acknowledge and honor past, present, and future Indigenous students of Chemeketa Community College.

Contents

Lab and Safety Regulations

Personal Behaviors

1. Eating and drinking are prohibited in the labs.
2. Wash your hands before leaving the lab.
3. Wash the lab counter & related work areas with disinfectant before and after the lab.
4. Stow your personal items in the cubbies or under your station.
5. Keep the lab counters and work areas uncluttered.
6. Clean-up is your responsibility:
    a. Return cleaned items back to lab kits.
    b. Wash slides and return to the original location (unless told otherwise). Cover slips may be thrown away.
    c. Wash glassware and return to original location (unless told otherwise).
    d. Ensure any lab waste is disposed of in the approved container(s).
    e. Ensure sinks and counters are as clean as when you arrived.
7. Clothing must be appropriate for the lab to be performed.
8. Read the lab in advance so you're aware of proper safety protocols.
9. If you have questions about safety, ask before you do.

Safety Protocols

1. Know where the lab safety equipment is located—fire extinguisher, first aid, eye wash, etc.
2. Wear personal safety equipment (goggles, gloves, aprons) as indicated in the lab instructions or by your instructor.
3. Handle all chemicals and biologicals, including stains, below eye level.
4. Do not assume something can just go down the sink or in the trash. Dispose of wastes in appropriate waste containers.
5. If there is a chemical spill:
    a. Get your instructor.
    b. If the spill is on the floor or counter, keep away from it.
    c. If the spill is on you or your clothes, rinse the area immediately with running water.
    d. If you splash chemicals in your eyes, go immediately to the eye wash station, turn on cold water, remove the red caps and lean down so that the water bubbles into your eyes.
6. In case of injury:
    a. Get your instructor.
    b. If the instructor is not available, call campus safety (x5023) on the lab's phone.
7. If the injury appears severe, also call 911
8. In case of fire:
    a. If your clothes are on fire, yell "FIRE" and roll on the floor or use a coat or fire blanket to smother it.
    b. If chemicals or lab equipment is on fire, call out "FIRE" and evacuate the room.
9. If the fire alarm goes off, take your essential items and leave the room. Head to the evacuation area as told by your instructor. Stay with your class and wait for instructions. Do not re-enter the building until an all-clear is given.

Name: ______________________ Lab Time: ______________ Due: ________________

1

1 Using the Scientific Method Pre-Lab

1. Consider each of the following statements in Table 1.1. Check the box next to the statement(s) that meet the requirements of a scientifically valid hypothesis.

Table 1.1. Scientific Hypothesis Statements

√	Statement
	I won the lottery because I am lucky.
	I received an A on my biology exam because I studied every day.
	My keys keep disappearing because there is a ghost in my house.
	No human lives forever.

2. Write a scientific hypothesis that would answer each of the following questions.

REMEMBER: If your hypothesis reads anything like an "if…then" statement or is focused on the results of an experiment, you probably provided a prediction rather than a hypothesis!

Why are the leaves on my houseplant turning yellow?

Why do I get pimples right before final exams?

3. Describe a control for each of the following experiments.

   A plant biologist wants to know how a particular plant hormone affects plant cell growth, so he treats different plants with different concentrations of the hormone.

   An ecologist is trying to determine the role of a species of sea star in a tide pool community, so he removes the species from a specific area of the tide pools.

4. Which of the experiments in Table 1.2 is the best way to include replicates in the experiment described? Check the box next to the experiment.

Table 1.2. Tomato Experiments

√	Experiment
	A tomato plant has 15 leaves. Five leaves are covered with blue filters, five with red filters, and 5 with black paper. The plant is exposed to bright light for 3 days, and then the leaves are tested to determine how much starch they produced.
	Fifteen tomato plants are pruned to contain 15 leaves. Five leaves on each plant are covered with blue filters, five with red filters, and 5 with black paper. The plants are exposed to bright light for 3 days, and then the leaves are tested to determine how much starch they produced.
	Three tomato plants are pruned to have 15 leaves. The leaves of one plant are covered with blue filters, the leaves of the second plant are covered with red filters, and the leaves of the third plant are covered with black paper. The plants are exposed to bright light for 3 days, and then the leaves are tested to determine how much starch they produced.

Explain your answer.

Controlled Experiment Description

To examine the effect of organic and synthetic fertilizers on seedling growth, fava beans (*Vicia faba*) were planted in unfertilized sand, sand amended with synthetic fertilizer, and sand amended with organic fertilizer. After germination, seedling height was measured daily.

The seedlings that were given different fertilizers grew faster than those that were unfertilized, but the rate of seedling growth differed little between the two types of fertilizer (Figure 1.1). The average seedling height 5 days after germination was 19 cm in the plants fertilized with organic fertilizer and 17 cm in the plants fertilized with synthetic fertilizer. Unfertilized plants grew to approximately 5 cm. The rate of growth is constant for the organic and synthetic fertilizers, as indicated by the similar slope of the lines for these treatments in Figure 1.1. Unfertilized plants initially grew at the same rate as fertilized plants but stopped growing after Day 3.

Figure 1.1. Seedling Growth in Soils With and Without Fertilizer

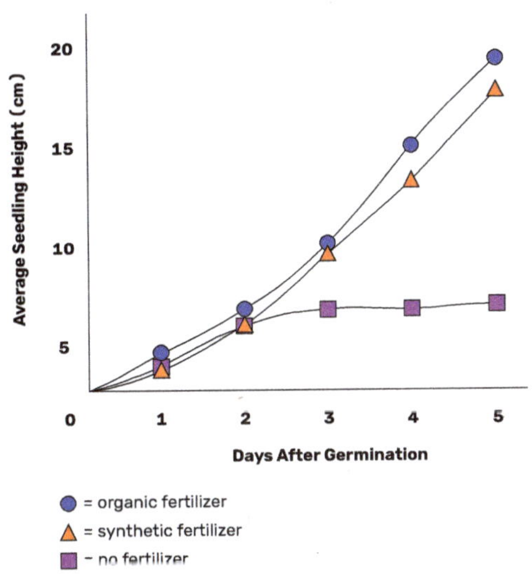

5. Use the **controlled experiment description** to answer the following questions.
   What question is the controlled experiment designed to answer?

   What are the dependent variables in the controlled experiment?

   What is the independent variable in the controlled experiment?

1

Activity 1.1. Applying the Scientific Method

During this lab, your group will design and run an experiment, collect data and analyze results. The subject of the experiment will be determined by your instructor and the supplies available in the classroom.

A. Observations

Watch the videos and examine the material provided by your instructor to learn more about the test subject. Record your observations on a piece of paper or white board as you work.

B. Form a question

Procedure A. Form a Question and Hypothesis

As a group, form a **question** about the test subject and propose a hypothesis. The steps of the scientific method can go in any order. If you have an idea for an experiment, you can start there and work backward to determine the question and hypothesis, but you must have all three. Choose carefully...

Write the group question on a whiteboard. After receiving instructor approval, record the question in Table 1.4

Instructions

1. Write the group question on the whiteboard with your question. Read the question and hypothesis together. Does the hypothesis answer the question? After receiving instructor approval, record the question in Table 1.4.

Check with your instructor to ensure that your group is asking an appropriate question.

Form a hypothesis in answer to your question.

Different members of your group may have different hypotheses.

Remember that a hypothesis is a possible answer to your question and not another question or an "if...then" statement.

2. Record your hypothesis in Table 1.4.

Check with your instructor to ensure that your group is asking an appropriate question.

Procedure B. Plan a Group Experiment

Form a hypothesis in answer to your question. Different group members may have different hypotheses.

Instructions

1. Write out step-by-step instructions for your experiment on a whiteboard.

Check with your instructor to ensure you have a controlled experiment.

2. List your variables (independent, dependent and at least 2 controlled) in Table 1.4.
3. Complete Table 1.4 by forming a prediction of what will happen in your experiment if your hypothesis is valid.
4. Design a data table (Data Table 1.1) for your experiment in the space provided. *See the results section of Appendix A to learn more about how to properly format a table in a scientific paper.*
5. Plan for at least three trials of your experiment.

Check with your instructor to ensure that you haven't missed anything.

Data Table 1.1. Title:

1

Procedure C. Run the Experiment

The development of an experiment requires that you define the variables in the experiment. A variable is any factor that might affect the outcome of an experiment. In this class, we will work primarily with controlled experiments, in which we change only one variable at a time and observe the outcome. The variable that we change, the one we measure, and all the ones that do not get changed have special names:

Dependent variable: This is the factor that will be measured in your experiment. It is considered "dependent" because the measurements "depend" on the treatment.

Independent variable: This is the factor that your hypothesis focuses on. It is the thing that changes between the different samples in your experiment.

Controlled variables: These are the factors that must be kept constant for all the samples in your experiment. When you design an experiment, you want to ensure that only your independent variable is affecting your results. All other factors must be kept the same between all the samples.

Instructions

1. Run the experiment and record the data in DATA TABLE 1.1.
2. Average the results (dependent variable) of the three trials.

Table 1.4. Notes on Experimental Design

Group Question	Instructor Initials __________
Your Hypothesis	Instructor Initials __________
Step-By-Step Procedure	**Complete on Whiteboard** Instructor Initials __________
Independent Variable	
Dependent Variable	
Control Treatment	
Controlled Variables	
Prediction	

Procedure E. Analyzing and Displaying Data

A table is not always the best way to analyze your data. Creating a graph often allows you to visualize trends that are not apparent in the numerical format of a table. *Consult the Results section of Appendix A to learn more about how to properly format a graph for a scientific paper.*

Instructions

1. Create the data table your group will use to gather data.
2. After gathering your data, create a graph to display the average of your trials in Figure 1.2.

Figure 1.3. Graph Title:

Procedure E. Communicating Your Data

Collaboration is key to the scientific process. Scientists present their results frequently to their own lab members and to a broader audience at conferences. Regardless of the audience, your peers critique your work, provide insights based on their personal background, and ask questions that help you delve deeper into your topic. If time allows, each group will present the results of their experiment to the class.

Instructions

1. Create a short presentation of your group's results. Assign one of the following roles to each of your group members:
   - ❑ Question, Hypothesis, and Reasoning
   - ❑ Experimental Design
   - ❑ Describe and Summarize Graph
   - ❑ Draw a Conclusion

1

Name: ____________ Lab Time: ____________ Due: ____________

1

1 Applying What You've Learned

1. Consider the data sets represented in Table 1.3. Which is continuous?

Table 1.3. Data Sets

√	Data Set
	The average temperature in Salem is recorded daily at 4 pm for a week.
	The number of Chemeketa students who claim their favorite class is biology, chemistry, or physics.

2. Consider each of the following data sets. Would it be better to use a table or a graph? Explain your reasoning.

High temperatures in Salem: October 1, 72°F; October 2, 75°F; October 3, 68°F, October 4, 69°F, October 5, 71°F

Favorite subjects of Chemeketa students: Biology, 93; Chemistry, 38; Physics, 37

3. Return to the two data sets from Question 6. Assume that you decided both would be better as a graph (there are a few possible answers on that one!). What type of graph would you use, bar or line? Explain your answer.

Temperature Graph

Favorite Subjects Graph

4. Use the following summary to properly format the graph in Figure 1.4. Add axis titles, a key, and a figure legend.

5. Every year, the 8th class at Average Middle School has 30 students. The number of 8th graders known to have a diagnosis of autism spectrum disorder (dashed line) was compared to the number of 8th graders who passed the school's exit exam (solid line) over a period of 10 years to determine if there was a correlation between having autism and success on the exam. In 2010, all 30 students passed the exit exam.

Figure 1.4. Graph Title:

Name: ______________________ Lab Time: ______________ Due: ______________

2 Microscopes & Micro-communities Pre-Lab

Activity 2.1. Learning About Microscopes

Visit the virtual microscope at http://go.chemeketa.edu/virtualmicroscope or by scanning the QR code. Use the “learn” button to explore the parts of the microscope. Figure 2.1 is a diagram of the microscope used in our classrooms.

1. Use what you learned from the virtual microscope to label the following parts of the microscope in Figure 2.1.

   a. Coarse focus knob
   b. Condenser
   c. Fine focus knob
   d. Iris diaphragm
   e. Mechanical stage adjustment knob
   f. Objectives
   g. Ocular
   h. On/off switch
   i. Rheostat
   j. Stage

Figure 2.1. Microscope Anatomy

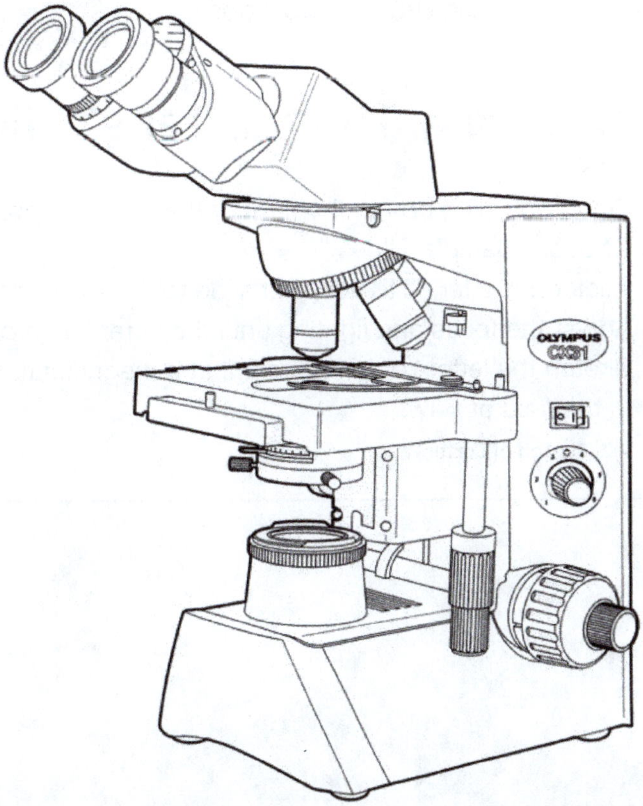

2. Indicate the part of the microscope that you would use in each of the scenarios in Table 2.1.

Table 2.1. Microscope Adjustment Scenarios

Scenario	Microscope Part to Adjust
The specimen is at the edge of your view when you look through the oculars, and you want to center it.	
The image through the oculars is slightly out of focus at 40× magnification.	
The image is in focus, but it is hard to see your specimen due to a lack of contrast.	
The light coming through the oculars is too bright for you to view the image comfortably.	
When preparing to store the microscope, you notice that the high-power objective is very close to touching the stage.	

Return to the virtual microscope. Use the "explore" button and the instructions below to practice viewing slides on the virtual microscope.

Procedure A. The "Letter e" Slide

Instructions

1. Select the slide box to the right of the microscope.
2. Choose "Sample Slides."
3. Click on "Letter e" to add that slide to the virtual microscope.
4. Adjust the focus and lighting until the letter "e" is clearly visible.
5. Sketch the letter "e" in Figure 2.2 at a magnification that allows you to see the entire letter in the field of view.

Figure 2.2. Sketch of Letter "e"

3. A common mistake for beginners is thinking that the letter "e" just disappears at high magnification. How would you explain to a younger student what happened to the letter "e" as you increase the magnification?

2

Procedure B. Plant Cells

Instructions

1. Return to the slide box.
2. Choose plants.
3. Select the "Plant Cells" slide.
4. Focus with 4×, 10×, and 40×.

You may need to adjust the light and center your slide.

4. Sketch three plant cells using the 40× objective.

5. What happened to the number of plant cells visible on your slide as you increased the magnification?

Procedure C. Human Blood

Instructions

1. Return to the slide box.
2. Choose "Human."
3. Select the "Blood" slide.
4. Focus with 4×, 10×, and 40×.

You may need to adjust the light and center your slide.

6. Sketch at least five blood cells using the 40× objective.

7. What happened to the amount of light available as you changed the magnification?

Name: ____________________ Lab Time: ____________ Due: ______________

2 Microscopes & Micro-Communities

Microscopes are commonly used by biologists whose research involves the study of very small specimens. Over the years, many types of microscopes have been developed. Some magnify an image to a small degree (1 time greater than you see with your eyes) while others magnify at much higher levels (more than 50,000 times greater than you can see with your eyes).

As microscopes have developed, so have a variety of preparation and staining techniques that enable scientists to extract a great deal of information from an image. In science, an image is worth much more than a thousand words.

The type of microscope and preparation technique you use depends on what you are looking at and what you want to see. This lab should familiarize you with the type of microscope most commonly found in classrooms: the **compound light microscope**. We will also take a few minutes to work with a **stereomicroscope**, also known as a **dissecting microscope**.

Learning Outcomes

- ❑ Identify and describe the function of the main parts of a compound light microscope and a stereomicroscope.
- ❑ Operate a compound light microscope and a stereoscope.
- ❑ Prepare a wet mount.
- ❑ Determine the size of a specimen viewed through a compound light microscope.
- ❑ Create a scaled sketch of organisms viewed with a compound light microscope.

Activity 2.2. Microscope Techniques

A variety of techniques are associated with the use of a microscope. Complete the following activities to familiarize yourself with these techniques.

When you look into a microscope, you see a circle of light called the **field of view**. As you increase magnification, the area of the specimen you can see without moving the slide becomes smaller. We say that the field of view *decreases* as magnification *increases*, and you can visualize this change in size by measuring the diameter of the field of view at each magnification with a ruler—just put a clear ruler on the stage (Figure 2.3). Since the objectives of your microscope don't change, the diameter of the field of view at each magnification should remain the same as long as you are using the same brand of microscope.

In this activity, we will measure the field of view at each magnification of our microscopes. Your measurements will be made in fractions of **millimeters** (mm) and converted to **micrometers** (µm). Micrometers (10^{-6} m) are the measurements used by microscopists.

Figure 2.3. Field of View vs. Diameter of View

Procedure A. Measuring the Field of View

Instructions

1. Examine the micrometer slide with the 4× objective on your microscope. You should see a tiny ruler through your oculars. That ruler is 2 mm long.
2. Use the mechanical stage adjustment knob to center the ruler in your field of view.
3. Estimate how many "rulers" it would take to cross the field of view you see with your 4× objective.
4. Multiply the number of rulers by 2 (the length of the micrometer in mm) to determine the diameter of the field of view in mm at 4×.
5. Enter your measurement into the appropriate cells of Table 2.2. Convert your measurement to micrometers to complete the row for the 4× objective.
6. Repeat this process for the 10× and 40× objectives. Remember that you are estimating!

Table 2.2. Field of View Size Using Different Objectives

Objective	Field of View Diameter (mm)	Field of View Diameter (µm)
4×		
10×		
40×		

Hint: Keep this table, or a version of it, with your lab notebook for future reference.

Procedure B. Preparing a Wet Mount

The slides most students are familiar with are **prepared slides**. To create a prepared slide, a specimen is preserved, sliced, and permanently mounted on the slide for use by multiple students. Prepared slides typically incorporate stains or dyes that highlight specific parts of the cell or tissue being examined. In the laboratory, you can make your own slides of fresh material by preparing a **wet mount**. Dyes can also be used to highlight cell structure in wet-mounted specimens.

2

Instructions

1. Obtain a clean slide and cover slip.
2. Place a drop of liquid in the center of the slide.
3. Place a small, thin specimen in the drop of liquid.
4. Hold the cover slip between your thumb and forefinger. Place one edge of the cover slip on the slide near the left side of the liquid containing your specimen. The cover slip should be at a 45° angle to the slide.
5. Slowly lower your cover slip onto the slide.
6. Tear a leaf off the water plant *Elodea* and follow the instructions above to prepare a wet mount as described. Take a moment to examine your wet mount with the microscope before proceeding.

Procedure C. Measuring a Specimen

Now that you know the field of view at each magnification of your microscope, you can use those measurements to estimate the size of the specimens you observe. Let's practice by determining the size of an *Elodea* cell.

Instructions

1. Focus on the specimen with the 4× objective.
2. Increase your magnification until you can clearly see the individual cells that make up your specimen. Under the microscope, the *Elodea* leaf looks like a green brick wall. Each brick in the wall is an individual cell.

> **Determining the best magnification to use when measuring your specimen.**
> Know what you are looking for! Most students want to start at the highest magnification and get stuck. Focus and center your specimen in the field of view with the 4× objective. Choose the lowest possible magnification that allows you to see the thing you are looking for.

3. Divide your lab notebook page into three sections by making two horizontal lines.
4. In the left half of the top third of your notebook, sketch a cluster of *Elodea* cells. Your sketch should fill the left half of that section. Don't draw teeny tiny cells, include internal details. Label any structures you can identify.
5. Record your written observations about your specimen in the space to the right of your sketch.

6. After you have finished sketching, return to the microscope to estimate the length of one cell in *Elodea* (one that is in your sketch) based on your measurements in Table 2.2. Jot that number down in the margins of this paper.

> *Hint:* Remember that you already know the diameter of your field of view. What proportion of that field of view does the cell occupy? For instance, if your field of view is 1000 μm and the cell occupies half of the diameter of the field of view, it is approximately 500 μm in size.

7. Create a scale bar for your drawing (See Figure 2.4 for an example).
    a. Count the number of grid squares that represent the length of that cell in your sketch.
    b. Divide your estimate of the cell size based on the microscope by the number of grid squares in your sketch to get a measure of the number of μm/grid square.
    c. Draw a scale bar based on grid squares and put the conversion factor above the scale bar. Remember that you are estimating here, so don't be afraid to round up or down. **The value of a scale bar should be 1 μm or a multiple of 5 μm.**

Figure 2.4. Creating a Scale Bar for Sketches

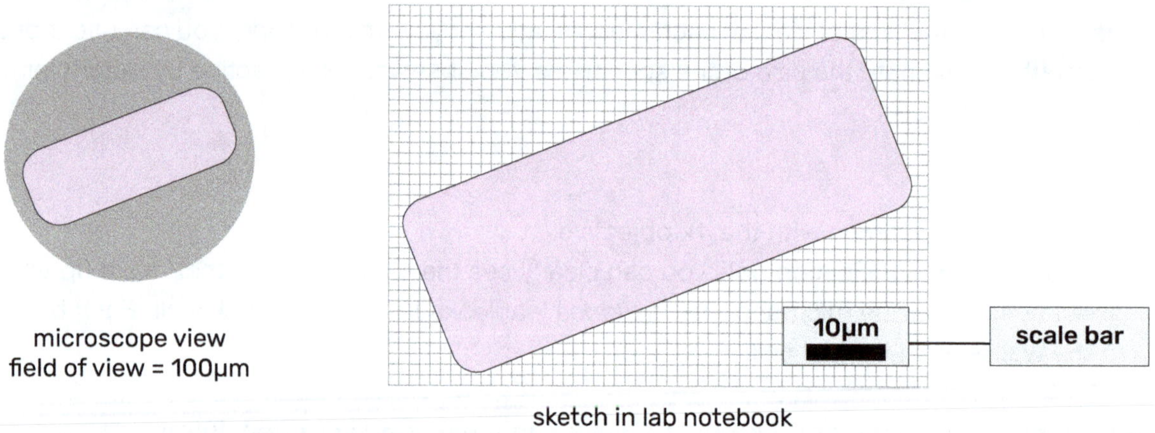

Q1. What happens to the size of the field of view as you increase magnification?

Q2. How do you determine the proper magnification to use when examining a specimen with the compound light microscope?

Q3. What magnification did you use to measure *Elodea* cells? Why was that magnification appropriate?

Activity 2.3. The Stereomicroscope

The stereomicroscope (also known as the dissecting microscope, Figure 2.5) behaves somewhat differently than the compound microscope. You will occasionally use this microscope during class, so it is essential to become familiar with its function as well.

Figure 2.5. Stereomicroscope

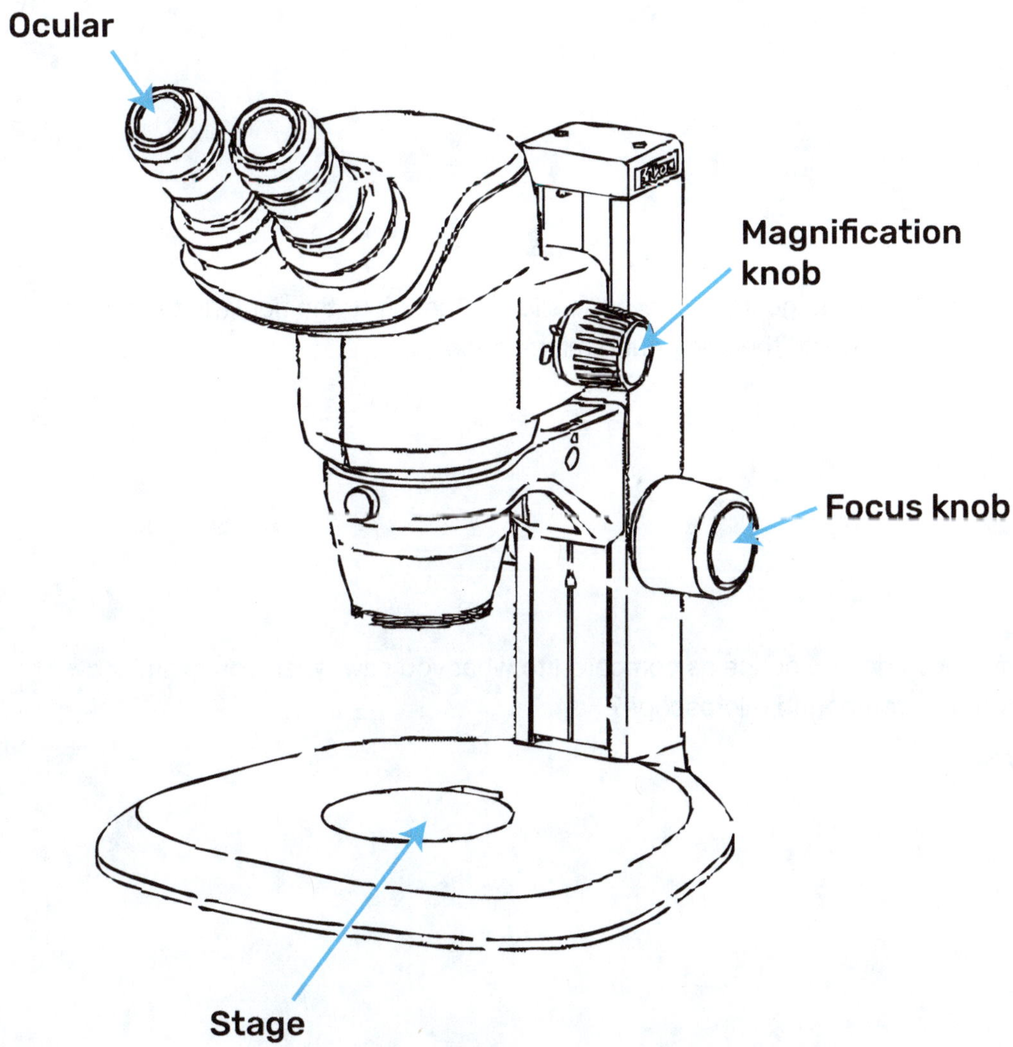

Procedure A. Identify the Stereomicroscope Parts

Instructions

1. Identify the following parts on the stereomicroscope
   - ❑ Oculars
   - ❑ Stage
   - ❑ Magnification knob (switches internal lenses to adjust the magnification)
   - ❑ Focus knob

2. Place the slide of letter "e" right side up on the stage and observe.
3. Examine the wet mount of *Elodea* on the stage of the stereomicroscope.

Q4. Does the stereomicroscope affect the orientation of the image in the same way that the compound microscope does? Explain your answer.

Q5. What happens to the image if you move the slide to the left or the right? Is the effect the same as you saw with the compound microscope?

Q6. What differences do you notice as compared to what you saw when you examined *Elodea* with the compound microscope?

Activity 2.4. Microscope Fun!

Time to have some fun and get more practice using the microscope. Your instructor has provided samples of moss and lichen soaked in water overnight. Moss and lichen are full of unseen critters that we will learn about over the term. For now, we will just observe samples and practice drawing and measuring the organisms we see.

Procedure A. Exploring Samples

Instructions

1. Obtain a sample from your instructor.
2. Lift the vegetation from the sample and gently squeeze it over the dish to dislodge any organisms that may be stuck to it.
3. Place the vegetation in the upturned lid of the sample jar.
4. Place the dish under the stereomicroscope.
5. Scan to find an area where movement is visible.
6. Use a pipette to obtain a sample from that area.
7. Add approximately three drops to the well of a depression slide and place a coverslip on top.
8. Place the slide on the compound microscope and examine with the 4× and 10× objectives.
9. Sketch at least three different organisms you observe to scale in your lab notebook using the procedure described in Activity 2.2.C. Record observations for each specimen.

> **REMEMBER**: Your notebook page should be divided into thirds, and the sketches and observations for one specimen should fill one-third of the page. That means no more than three specimens per page.

Activity 2.5. Storing the Microscope

Don't forget to clean up. Microscopes require special care when being stored.

Instructions

1. Rotate the nosepiece until the shortest (scanning) objective is in place and lower the stage to its lowest point.
2. Remove the slide from the stage.
    a. Clean it.
    b. Dry it.
    c. Put it back.
3. Clean off any water on the stage with a paper towel or absorbent paper.
4. Coil the cord on the back of the microscope.
5. Put the dust cover on.
6. Return the microscope to the cabinet with its oculars facing away from you.

Name: ______________________________ Lab Time: __________________ Due: ____________________

2 Applying What You've Learned

A selection of these questions will appear on your lab quiz. Lab quizzes are open-book. Type up answers to these questions in advance so that you can copy and paste the answers into the quiz.

1. Use your calculations of the diameter of the field of view in your class microscope to estimate the length (longest axis) of each organism shown in Figure 2.6 (along its longest axis). Enter that number in Table 2.3 below each image.

Figure 2.6. Measuring Organisms

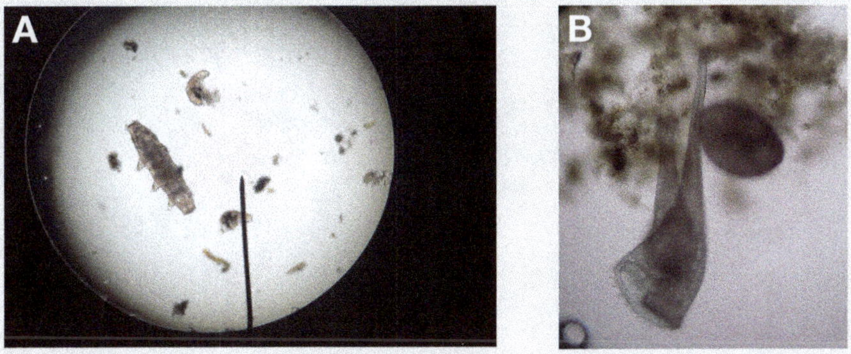

Table 2.3. Estimating the Length of Organisms

Objective Magnification (A)	40×
Estimated Size of Organism (A)	

Objective Magnification (B)	10×
Estimated Size of Organism (B)	

2. Describe **three** ways that a stereo microscope differs from a compound light microscope.

3. Provide an example of a situation where you would need to view a specimen with a stereomicroscope.

2

4. Provide an example of a situation where you would need to view a specimen with a compound microscope.

5. You used estimation in creating scale bars for your drawings. Is this an ethical practice? Would you publish a paper based on such estimates?

Name: ______________________ Lab Time: ______________ Due: ______________

3 Protistan Diversity Pre-Lab

Work with your group to answer the following questions. Collect your sample as close as possible to your lab day and time. Bring your group sample to class.

1. What factor(s) affect freshwater plankton diversity?

2. Explain why you think these factors are important to diversity.

3. Where will your group collect its sample?

4. Why did you choose this location (relate to hypothesis)?

5. Do you predict that your sample will exhibit high or low plankton diversity? Why?

6. Write a one-paragraph description of your collection site (to be shared with the class).

7. Submit three photographs of your collection site with Pre-Lab 3.

   a. Overview of the entire site
   b. Image of the area near the sample location (including vegetation)
   c. Image of sample location

Name: ______________________ Lab Time: ______________ Due: ________________

3 Protistan Diversity

The first **eukaryotic** organism evolved more than 2 billion years ago and was most likely a single-celled chemoheterotroph with few organelles (maybe just a nucleus). From there, eukaryotic life evolved to encompass a vast variety of life forms that inhabit diverse environments around the planet. Most of these organisms have been classified into the kingdom **Protista**. In this lab, we will learn to determine the clade (supergroup) of a protist as we explore the diversity of protists in our local environment.

Learning Outcomes

- ❑ Differentiate between a protist and the non-protist organisms found in a pond sample.
- ❑ Classify a protist at the level of supergroup.
- ❑ Examine the diversity of protists within various communities.
- ❑ Describe at least one factor that promotes protist diversity.

Activity 3.1. Protist Identification

In preparation for this lab, your group chose a location that you thought would exhibit a high diversity of protists and collected a sample with a plankton net. You will spend most of this lab analyzing your group sample. Your first step is to learn to distinguish between protists, plants, and animals. Second, you will use a simple key to sort your protists into one of the protistan supergroups.

Procedure A. Identifying Protists

Instructions

1. Allow the sample to settle for at least 15 minutes before examination.
2. While you wait, divide two pages in your lab notebook into six squares. Remember to use the squares on the left to make your sketches and the squares on the right to record observations.
3. Use a plastic pipette to place a drop of your sample on a slide. If you find that you have larger organisms, consider switching to a depression slide for your next sample.
4. Mount a coverslip over your droplet.
5. Find and create a scaled sketch of 6 different protists in your lab notebook.
6. Record observations of structure and behavior.

Activity 3.3. Diversity Calculations

3

To assess the diversity of a community, ecologists focus on two factors: species richness and relative abundance. **Species richness** refers to the total number of species present in a particular community and is a basic measure of community structure. **Relative abundance** refers to the number of individuals present for each species. Sometimes called **evenness**, relative abundance compares the population sizes of the species present in a given area.

A community can have high species richness, but if that community is dominated by one or a few species, it is not considered particularly diverse. A truly diverse community has approximately equal numbers of many different species. Due to the complexity of integrating both species richness and relative abundance into a single measurement, ecologists use mathematical models to combine these factors and obtain a single number that represents an index of the overall **diversity** of a community.

In this activity, we will utilize **Simpson's Diversity Index** to evaluate a range of communities. Note that other scientists use a variety of other mathematical models to assess species diversity. Each model has its own pros and cons. Simpson's Diversity Index is calculated with the following formula:

$$D = 1 - \sum n(n-1)/N(N-1)$$

where:

N = total number of individuals of all species

n = total number of individuals of a specific species

The result is a number that falls on a scale of 0 to 1 where 0 = lowest diversity and 1 = highest diversity.

Procedure A. Using Simpson's Diversity Index

1. Use Table 3.2 and your group's total data in Table 3.1 to calculate Simpson's diversity index for your group's sample.
2. Add your group's diversity index to the class table (Table 3.3).
3. Complete the Protist Diversity Group Assignment (available on Canvas) by forming a scientific explanation based on the class results that answers the question, "What environmental factor(s) affect plankton diversity?"

Table 3.2. Simpson's Diversity Index Calculations for ______________________ (location)

Supergroup	Number of Individuals (n)	n(n-1)		n(n-1)/N(N-1)
Archaeplastida				
SAR (Stramenopile)				
SAR (Alveolata)				
Excavata				
Unikonta				
Total Individuals (N) =		N(N-1)=	∑ n(n-1)/N(N-1)	
		D = n(n-1)/N(N-1) =		

Table 3.3. Simpson's Diversity Index at Various Salem Area Locations

Location	Simpson's Diversity Index

Name: ______________________ Lab Time: ______________ Due: ______________

3 Applying What You've Learned

A selection of these questions will appear on your lab quiz. Lab quizzes are open-book. Type up answers to these questions in advance so that you can copy and paste the answers into the quiz.

1. How do you know that an organism in your sample is a protist and not just a tiny plant or animal?

2. You probably saw both flagellate and ciliate protists in your sample (or watched a movie about it). How does the movement of these two types of organisms differ?

3. Which were more abundant in your plankton sample, heterotrophs or autotrophs? Why do you think this is the case?

4. The classification of protists is one of the most controversial areas of phylogenetics. Use the specimens you observed in this lab to explain why protists are so difficult to classify.

5. Students often assume that multicellular organisms are more successful than single-celled organisms. What's wrong with this assumption?

Name: ____________________ Lab Time: ____________ Due: ______________

4 Ecosystems & Energy Flow Pre-Lab

Scan the QR codes or go to the URLs to watch one video on owl pellets and one on biomass pyramids before answering these questions. You should conduct a small amount of additional research to answer some questions.

Owl Pellet Video (http://go.chemeketa.edu/owlpellet)

Biomass Pyramid Video (http://go.chemeketa.edu/foodwebs)

1. What is an owl pellet?

2. How many owl pellets does a barn owl produce per day? How many prey items are usually found in a barn owl pellet?

3. Differentiate between generalist predators and specialist predators.

4. Do you think an owl represents a generalist or specialist predator? Explain your reasoning.

4

5. Examine the dichotomous key to mammalian skeletons provided for this assignment.

   What trait will help you distinguish between an insectivore and a rodent?

   Describe several traits that will help you distinguish between the various “mouse” species and voles.

6. Assuming that owls consume all the animals on the dichotomous key, construct a hypothetical food web for the barn owl. You may have to do a bit of research to figure out what the prey species consume, but don’t spend too much time on this. Create a diagram of the food web of a barn owl. *Remember: A food web is not the same as a food chain! Arrows always point in the direction of energy flow.*

7. What is the difference between an energy pyramid and a biomass pyramid?

8. What is the connection between these two types of pyramids?

Name: ______________________ Lab Time: ______________ Due: ________________

4 Ecosystems & Energy Flow

Predators play a crucial role in the ecosystem, as their populations regulate the populations of lower-level consumers. Since predator populations tend to be smaller and easier to track than prey populations, they can be used to make inferences about prey populations and the health of an ecosystem.

We will focus on the Barn owls (*Tyto alba*), because they don't have to be trapped or dissected for us to see what they are eating. Barn owls consume their food whole and, after digesting and absorbing the nutrients, regurgitate any indigestible parts (bones, fur, feathers, etc.). The resulting **owl pellets** can be dissected, and the remains within them can be identified to determine what the owl is eating.

Learning Outcomes

- ❑ Use owl pellet data to estimate the biomass of different trophic levels of the owl food chain.
- ❑ Construct a biomass pyramid.
- ❑ Differentiate between generalist and specialist predators based on population growth curves.
- ❑ Use published data to interpret the food habits of owls.

Activity 4.1. Owl Pellet Dissection

Owl pellets can contain many things, but most of the ones we will examine will contain skulls. Any skull with teeth is a small mammal, and the teeth (size, shape, arrangement) are key to their identification. The skulls you will examine most likely belonged to a vole, mouse, rat (rodents) or a shrew (insectivore).

> *Important Information*
> - The owl pellets used in this lab are sterilized before being sent to us. Don't touch "wild" pellets without gloves, as they may contain harmful bacteria or fungi.
> - Be sure to count each individual only once. Mammal skulls come with a cranium (often crushed in the owl pellet) and a jaw. Jaw bones come in pairs (one pair = one prey animal). See Figure 4.1.

Figure 4.1. Contents of One Dissected *Tyto alba* Pellet

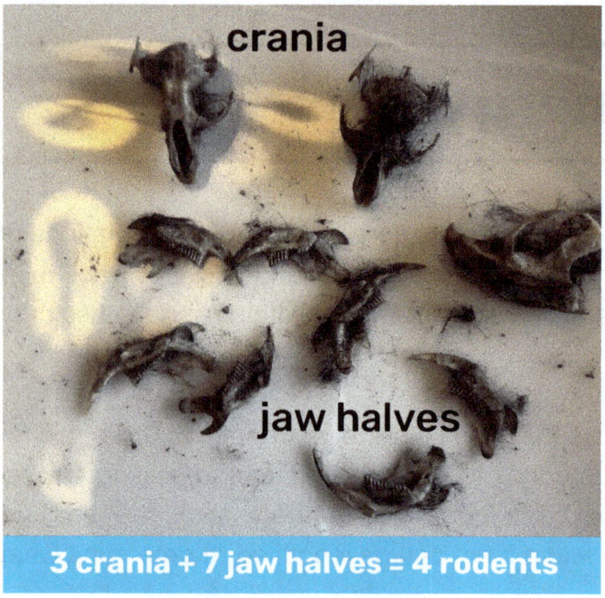

Instructions

1. Unwrap one owl pellet and place the pellet on your work surface.
2. Use your fingers to gently pull the pellet apart.
    a. Tweezers and toothpicks can aid in dissection.
    b. If the pellet is too hard to dissect, you may add a spray of water, but be forewarned that the water will make it smell.
3. Separate bones for fur and skulls from jaws.
4. Discard all non-bone debris.
5. Repeat dissection with the second pellet.
6. Wash your hands and any equipment used with soapy water.
7. Use the dichotomous key (Figure 4.2) to identify the number and type of species in your pellet.
    a. Differentiating between species of small mammals can be a challenging task.
    b. For this activity, we will group our organisms into major groups (e.g., voles, mice, shrews) rather than identifying them to the species level.
8. Record the number of each type of prey species found in pellet #1 in the appropriate cells of Table 4.1. Record the same information for pellet #2 in Table 4.2.

Figure 4.2. Owl Pellet Dichotomous Key

Owl Pellet Dichotomous Key

1a. No teeth	BIRD (Figure 1)
1b. Teeth	Go to 2
2a. Teeth pointy, no gap between cheek teeth and incisors	Go to 3
2b. Teeth flat, gap between cheek teeth and incisors (Figure 2)	Go to 4
3a. No gap, zygomatic arch complete, skull flat and broad	MOLE
3b. No gap, zygomatic arch NOT complete, skull NOT flat and broad	SHREW
4a. Upper incisors grooved	MOUSE
4b. Upper incisors not grooved	Go to 5
5a. No infraorbital canal	POCKET GOPHER
5b. Infraorbital canal present (Figure 3)	Go to 6
6a. Angular cheek teeth that see to run together (Figure 4)	VOLE
6b. Lobed cheek teeth that are distinct	RAT

Figure 1. Bird Bones from an owl pellet

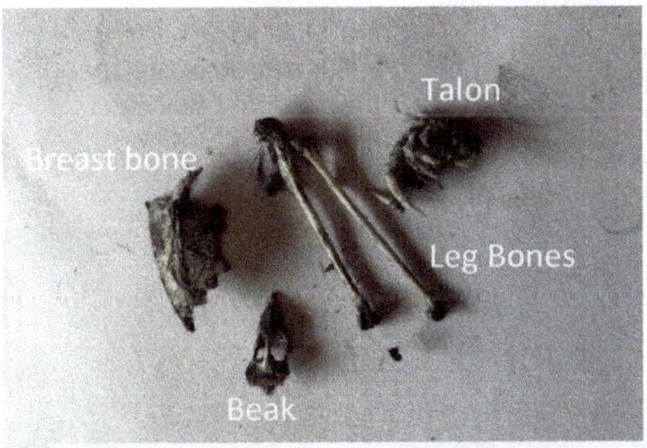

Figure 2. Vole jaw

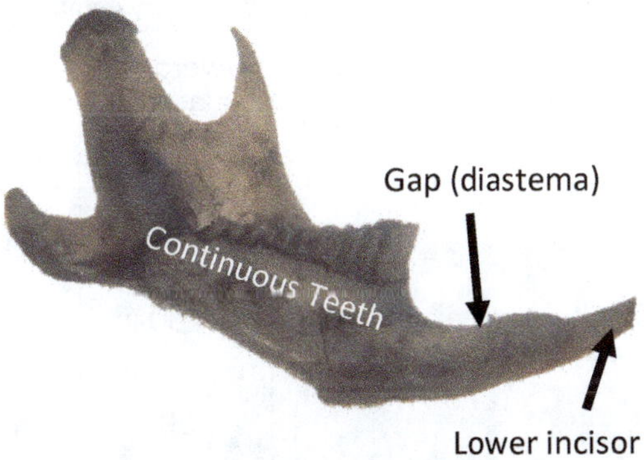

Figure 3. Vole (left) and pocket gopher (right) crania.

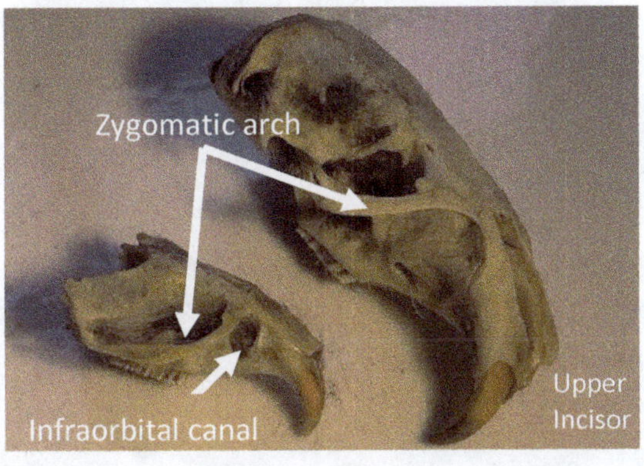

Figure 4. Vole jaw (left) with cheek teeth that appear to run together. Pocket gopher jaw (right) with distinct cheek teeth.

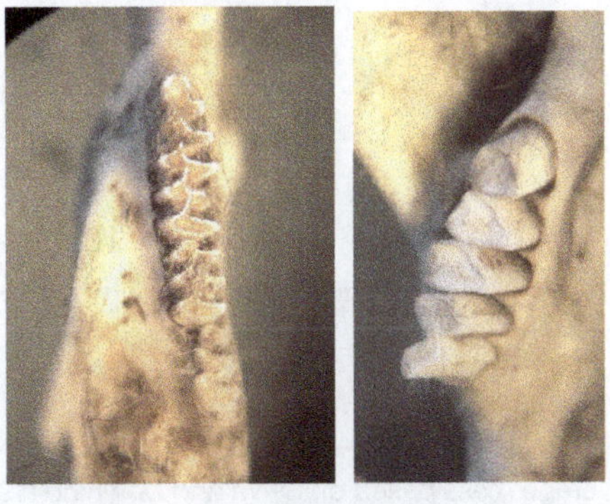

Table 4.1. Estimated Biomass of Prey Consumed by *Tyto alba* per Day (Owl Pellet #1). Occurrence refers to your chance of finding that organism in an owl pellet. ***** = very common, *** = common, ** = occasional, * = rare

Prey Species	Occurrence	Number in pellet	Average mass of species (g)	Live biomass in pellet (g) *# in pellet × average mass*	Live biomass consumed / day (g) *live biomass × 1.4*
Pocket Gopher	***		120		
Rat	*		250		
Mole	*		50		
Vole	*****		35		
Mouse	****		25		
Shrew	**		15		
Bird	**		10		
				Total live biomass consumed by owl per day (g)	

Data for average mass based on Sylva and Downing (1995).

Table 4.2. Estimated Biomass of Prey Consumed by *Tyto alba* per Day (Owl Pellet #2). Occurrence refers to your chance of finding that organism in an owl pellet. ***** = very common, *** = common, ** = occasional, * = rare

Prey Species	Occurrence	Number in pellet	Average mass of species (g)	Live biomass in pellet (g) *# in pellet × average mass*	Live biomass consumed / day (g) *live biomass × 1.4*
Pocket Gopher	***		120		
Rat	*		250		
Mole	*		50		
Vole	*****		35		
Mouse	****		25		
Shrew	**		15		
Bird	**		10		
				Total live biomass consumed by owl per day (g)	

Data for average mass based on Sylva and Downing (1995).

Activity 4.2. Prey Biomass & Energy Determination

Each individual prey consumed by an owl represents a specific amount of energy. That energy is captured through the consumption of producers (or consumers, when discussing shrews) and stored in the tissues of the individual. One way to measure the energy stored in an organism is to determine its **biomass** (weight of biological material). The biomass of any given animal is determined by a variety of factors, including genetics, age, and diet, and can vary significantly.

4

Instructions

1. Calculate the biomass of producers consumed by prey species in one of your pellets by multiplying the number of individuals of the prey species found in your pellets by the average mass of producers consumed by that species.
   a. Enter your calculations in Table 4.3 (owl pellet #1).
2. On average, a barn owl produces 1.4 pellets per day (Higuchi & Abe, 1997). To estimate the biomass of producers consumed by the prey of your owl for an entire day, multiply the biomass of producers consumed by prey species in your pellet by 1.4 (Table 4.3, column 5).
   a. Add the values in column 5 together to determine the total biomass of producers required to support your owl's prey for one day. Enter this total in Table 4.5, column 2.
3. The producers in a standard field contain, on average, 4 kcal of energy per gram of biomass (Golley, 1960). To estimate the energy of the producers consumed by the prey of your owl per day, multiply the producer biomass consumed by each prey species (g) by 4 kcal/g (Table 4.3, column 6).
   a. Add the values in column 6 together to determine the total energy requirements for each prey species in your pellet to determine the total estimated energy of producers consumed by prey per day.
4. Repeat for your second owl pellet using Table 4.4.
5. Complete calculations in Table 4.5 to find the total amount of energy consumed (kCal) by each owl per day.

Table 4.3. Daily Producer Consumption by *Tyto alba* Prey Species (Owl Pellet #1)

Prey Species	Number in pellet *(from Table 4.1)*	Average biomass of producers consumed by prey per day (g)	Biomass of producers consumed by prey (g) *# in pellet × biomass consumed*	Estimated producer biomass consumed per day by prey (g) *mass of producers × 1.4*	Estimated energy of producers consumed by prey (kcal) *biomass consumed × 4 kcal/g*
Pocket Gopher		90			
Rat		53			
Mole		1000			
Vole		40			
Mouse		3			
Shrew		3200			
Bird		348			
			Total		

Producers are assumed to contain about 4 kcal of energy per g of biomass (Golley, 1960).

Table 4.4. Daily Producer Consumption by *Tyto alba* Prey Species (Owl Pellet #2)

Prey Species	Number in pellet *(from Table 4.1)*	Average biomass of producers consumed by prey per day (g)	Biomass of producers consumed by prey (g) *# in pellet × biomass consumed*	Estimated producer biomass consumed per day by prey (g) *mass of producers × 1.4*	Estimated energy of producers consumed by prey (kcal) *biomass consumed × 4 kcal/g*
Pocket Gopher		90			
Rat		53			
Mole		1000			
Vole		40			
Mouse		3			
Shrew		3200			
Bird		348			
			Total		

Producers are assumed to contain about 4 kcal of energy per g of biomass (Golley, 1960).

Table 4.5. Estimated Energy Consumption of *Tyto alba* per Day

	Live biomass consumed/day (g) *(total from Tables 4.1 and 4.2)*	Available energy *kCal / g*	Energy consumed by owl (kCal) / day *biomass/day × energy*
Owl Pellet #1		1.58	
Owl Pellet #2		1.58	
Available energy represents the average amount of energy stored in each gram of tissue for common owl prey (Fleharty et al, 1973).			

4

Activity 4.3. Producer Biomass & Energy Determination

In 1957, ecologist Howard Odum published a classic paper demonstrating how the second law of thermodynamics affected the efficiency of energy transfer between trophic levels. One way to examine the flow of energy through an ecosystem is to use data to create energy pyramids. Generally, only 5–20% of the energy at one trophic level is transferred to the next. This work led to the "10% rule," which states that on average, only 10% of energy is transferred between trophic levels.

Research in science always requires replication. In this case, we are looking at owl pellets from a variety of owls (two owls for each member of your class). Owls, like humans, vary in size, shape, and even in hunting ability. Just like us, they have good days and bad days. All of this creates a good deal of variation in the data. For this reason, it is better to look at the averaged data from a large number of specimens rather than the data from just one owl. We can use that average to estimate the number of producers (in terms of biomass and energy) required to support the prey species for one owl.

Instructions

1. Use the link provided on Canvas to open your class' Owl Pellet Database. Type directly into the database. Please do not change data entered by other students!
    a. Add your initials to the "student initials" column (once for each owl pellet you dissected). Enter the data for each of your owl pellets on different rows.
    b. Add the following data points to from your data tables to the class data table (you'll find these values in the shaded areas in cited tables):
        i. total live biomass consumed per day (Tables 4.1 and 4.2)
        ii. the total producer biomass consumed by prey species per day (Tables 4.3 and 4.4, column 5)
        iii. the total energy of producers consumed per day by prey species (Tables 4.3 and 4.4, column 6).
        iv. the total energy consumed by the owl per day (Table 4.5)

c. The average per day and the average per year for each column is automatically calculated at the bottom of the spreadsheet.

d. Please wait to proceed until everyone in the class has entered their data.

2. Add the averages from the class database to the appropriate cells of Table 4.6.
3. To determine the energy available to prey species, divide the average annual energy consumption of the prey species by 0.10. Record your answer in the appropriate cell of Table 4.7.
4. Biomass also follows the 10% rule. To determine the biomass available to the prey species, divide the average daily biomass of producers consumed by 0.10. Record your answer in the appropriate cell of Table 4.7.

4

Table 4.6. Average *Tyto alba* Biomass and Energy Consumption per Year

	Average biomass consumed per year	Average energy consumed per year
Producers		
Primary & Secondary Consumers		

Compilation of data from Tables 4.1–4.5 and class database.

Table 4.7. Biomass and Energy Estimations for *Tyto alba* Food Chain

	Average biomass available (g) per year *average biomass consumed per year / 0.1*	Average energy available (kCal) per year *average kcal consumed per year / 0.1*
Producers		
Primary & Secondary Consumers		
Owl	250	1392.5

This is an estimate of how much energy is available in each based on the 10% rule (Odum et al., 1965). The last row is completed for you based on the following: the average biomass of an owl is 250 g and the average non-migratory bird contains 5.57 kcal/g of energy (Odum et al., 1965).

Activity 4.4. Biomass & Energy Pyramids

Energy (or biomass) pyramids are used to graphically display the energy flow through a food chain. Each trophic level is assigned a bar representing the amount of energy (or biomass) that exists at each trophic level. The loss of energy (and therefore biomass) between each trophic level means that the bar for each level is successively smaller. The final product resembles a pyramid, and the steps of the pyramid should illustrate a 90% loss of energy at each level. For instance, in the pyramid in Figure 4.3, the producers have 1000 kcal of energy. Each square represents 100 kcal (10 squares × 100 = 1000 kcal).

Figure 4.3. Sample Scaled Energy Pyramid. Each complete square in the pyramid (see key) equals 100 kcal.

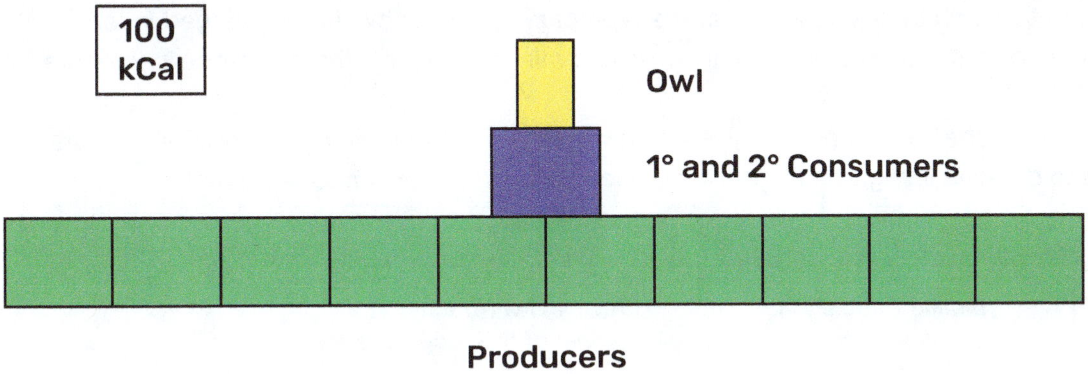

Procedure A. Creating a Biomass & Energy Pyramid

Use the data in Table 4.6 to create **both** an energy pyramid **and** a biomass pyramid for a barn owl food chain.

Instructions

1. Create a key.
    a. In your lab notebook, assign a value to a reasonably sized square (usually a 2 × 2 cell on graph paper). Set that square to be equal to a specific number (multiple of 5) of units (g for biomass or kcal for energy).

What value? That depends on your data. Looking at the figure above, one square equaled 100 kcal. The producers are represented by 10 squares or 10,000 kcal. You do not want a value that is so big that the producer bar does not fit on a single sheet of paper.

Notice also that the owl bar is less than one square. You don't want your value for a square to be so low that you cannot accurately represent the owl in your pyramid. Ask for help if you need it!

2. Draw a horizontal bar (2 squares tall and as long as your key value requires) to represent the amount of biomass/energy you calculated that producers needed to have to support the average owl.
3. On top of that, draw a bar (centered over the producer bar) to represent the amount of biomass/energy you calculated that the owl prey species needed to have to support the average owl.
4. On top of that (and centered), draw a bar representing the biomass/energy found in the owl.

4

Activity 4.5. Generalist or Specialist?

Barn owl pellets have been studied for many years, and much is known about their dietary preferences. Let's begin by taking a look at the type of predator an owl is. We learned in lecture that predators can be specialists or generalists (take a minute to review the difference in your pre-lab).

Table 4.8 shows the populations of mice and tawny owls in England over a period of 11 years. Use the data to create a line graph (Figure 4.4). Use a different color line to represent owls and mice.

Table 4.8. Tawny Owl and Mice Populations in England from 1949–1960 (after Southern, 1970)

Year	Mouse Population *(mice / acre)*	Tawny Owl Population *(number of breeding pairs)*
1949	Spring: 6; Fall: 3	19
1950	Spring: 18; Fall: 7	20
1951	Spring: 9; Fall: 5	20
1952	Spring: 6; Fall: 12	20
1953	Spring: 6; Fall: 2	21
1954	Spring: 5; Fall: 3	21
1955	Spring: 19; Fall: 2	22
1956	Spring: 12; Fall: 6	23
1957	Spring: 9; Fall: 15	24
1958	Spring: 10; Fall: 1	24
1959	Spring: 8; Fall: 12	24

Figure 4.4. Owl and Mice Populations in England (1949–1960)

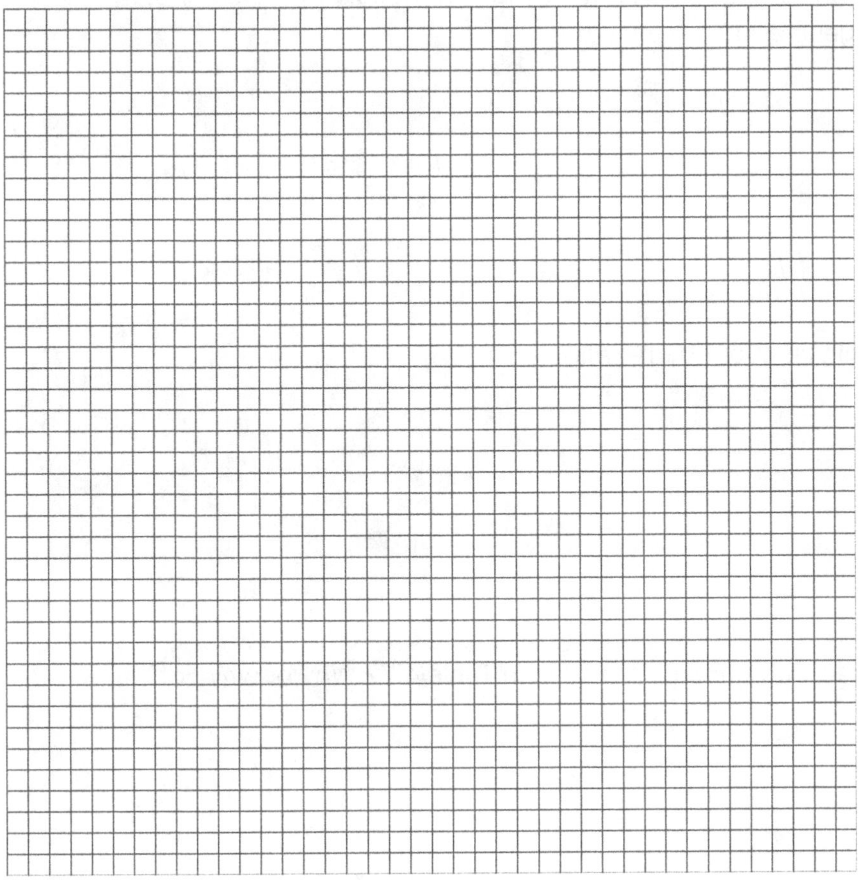

Q1. Are the owls in this study generalists or specialists? Use your graph in Figure 4.4 to explain your reasoning.

Q2. What would the graph of the other type of predator (generalist/specialist) look like? Explain.

4

Q3. Table 4.8's data has provided us with additional information about the diet of tawny owls, but does it align with what we observed in our owl population? Do our data suggest the barn owl is a generalist?

Q4. What does this data tell you about barn owls?

Q5. Do these data sets agree with the data provided in Table 4.7? Why or why not?

Q6. Which of these data sets gives us more information about the dietary habits of the barn owl? Explain your reasoning.

Q7. In both of these data sets, barn owls eat a significantly higher number of voles than other species. Is this because voles are more common in the population? Are there other possible explanations?

4

Q8. Based on these data sets, what can we say about the diet of the barn owl? Is the answer as straightforward as stating that the owl is a generalist or specialist? Explain your reasoning.

Q9. Is there additional information not provided with these datasets that would provide insight as to the diet of the barn owl? Describe one example.

Name: ____________________ Lab Time: ______________ Due: ______________

4 Applying What You've Learned

A selection of these questions will appear on your lab quiz. Lab quizzes are open-book. Type up answers to these questions in advance so that you can copy and paste the answers into the quiz.

1. Do either of your pyramids support the Odum's 10% rule? Why or why not?

2. Compare the biomass pyramid and the energy pyramid. Do these two pyramids provide you with the same information? Explain your answer.

3. Which type of pyramid is more practical to build in the field? Which is more accurate?

4. In this activity, we combined the primary and secondary consumers into a single trophic level. How would your pyramids change if we were to calculate their contributions separately?

5. If the diet of an owl were exclusively based on shrews, would you expect the number of skulls in your pellets to be greater or less than the number of skulls found in the pellets of owls that consumed a variety of prey? Explain your answer.

References

Fleharty, EM, Krause, M, and Stinnett, D (1973) Body composition, energy content, and lipid cycles of four species of rodents. *J. Mammal*. 54(2): 426–438.

Golley, FB (1960) Energy dynamics of a food chain in an old-field community. *Ecol. Monogr.* 30(2): 187–206.

Higuchi, A. and M. Abe (1997) Estimation of food consumption from pellets cast by captive Ural owls (*Strix uralensis*). In *Biology and Conservation of Owls of the Northern Hemisphere: 2nd International Symposium* (Duncan, J et al. Eds). General Technical Report NC-190. St. Paul, MN:US Dept of Agriculture, Forest Service, North Central Forest Experiment Station.

Maser C, Shaver S, Shaver C, Price B (1980) A note on the food habits of the Barn Owl in Malheur County, Oregon. *The Murrelet* 61 (2): 78–80.

Odum, E., S. Marshall, and T. Marples (1965). The caloric content of migrating birds. *Ecology*, 46(6):901–904.

Pearson O and Pearson A (1947) Owl Predation in Pennsylvania, with notes on the small mammals of Delaware County. *J. Mammalogy* 28 (2): 137–147.

Sylva, M, and Downing, J (1995). *Handbook of Mammalian Body Masses*. CRC Press, Boca Raton, FL.

Southern H (1970) The natural control of a population of tawny owls (*Strix aluco*). *J. Zool. Lond.* 162:197–285.

4

Name: ______________________ Lab Time: ______________ Due: ______________

5 Oregon Coast Phylogenetics Pre-Lab

Introduction

To prepare for this week's lab, you will work through the first two "missions" of Nova's *Evolution Lab* at https://go.chemeketa.edu/evolutionlab (or at the QR code link) to get a better understanding of how evolutionary trees are created and used. Complete this handout as you work. In some cases, the questions help you sum things up, and in others, they will help you figure out what to do next.

5

Before beginning Puzzle 5.A, watch the introductory video to learn some tree basics and to get an overview of how the *Build A Tree Virtual Lab* works, and answer the following questions using Figure 5.1.

Figure 5.1. Evolutionary Phylogenetic Tree

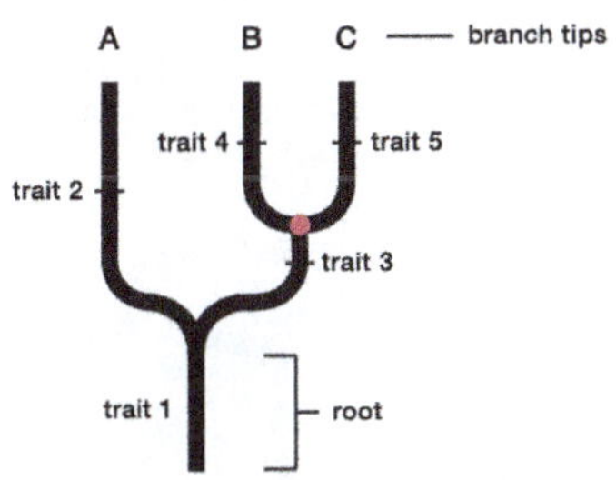

1. What does the circled node represent? *Check the box next to the correct answer.*
   - ❑ The common ancestor species of B and C
   - ❑ A speciation event
   - ❑ Both a and b

2. Which way does time run on this tree? *Check the box next to the correct answer.*
   - ❑ From the root to the branch tip
   - ❑ Across branch tips, from left to right

3. Which lived more recently in time? *Check the box next to the correct answer.*
   - ❑ Common ancestor species of A and B
   - ❑ Common ancestor species of B and C

4. Which traits do A and B share? *Check the box next to the correct answer.*
   - ❑ Trait 1
   - ❑ Trait 2
   - ❑ Trait 3
   - ❑ Trait 4
   - ❑ Trait 5

5. Which traits do B and C share? *Check the box next to the correct answer.*
   - ❑ Trait 1
   - ❑ Trait 2
   - ❑ Trait 3
   - ❑ Trait 4
   - ❑ Trait 5

6. Use your answers to questions 3–5 to explain why B and C are more closely related to each other than A and C are.

7. There is more than one correct way to show relationships using a phylogenetic tree. Which of the trees in Figure 5.2 shows the same exact relationships as the tree in Figure 5.1? You may circle more than one.

Figure 5.2. Showing Relationships Using Phylogenetic Trees

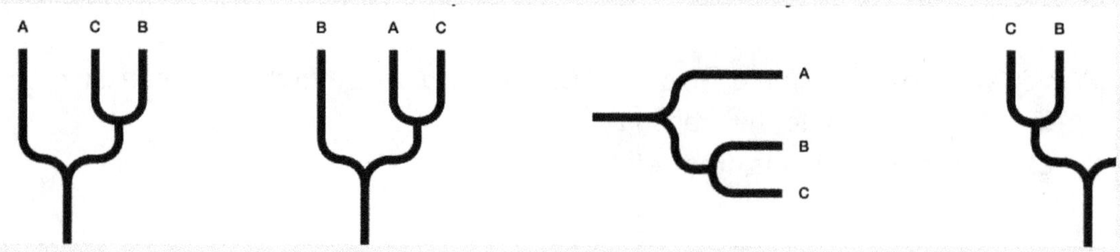

Understanding Phylogenetic Trees

Puzzle 5.A. Red, Green, and Gecko

8. Under a microscope, the cells of mushrooms, plants, and animals all have visible nuclei. This makes them all what? *Check the box next to the correct answer.*
   - ❑ Autotrophs
   - ❑ Heterotrophs
   - ❑ Eukaryotic
   - ❑ Prokaryotic

9. What trait do the mushroom and gecko share that the tree lacks?

Puzzle 5.B. Familiar Faces

Figure 5.3. A Phylogenetic Tree With Amniotes

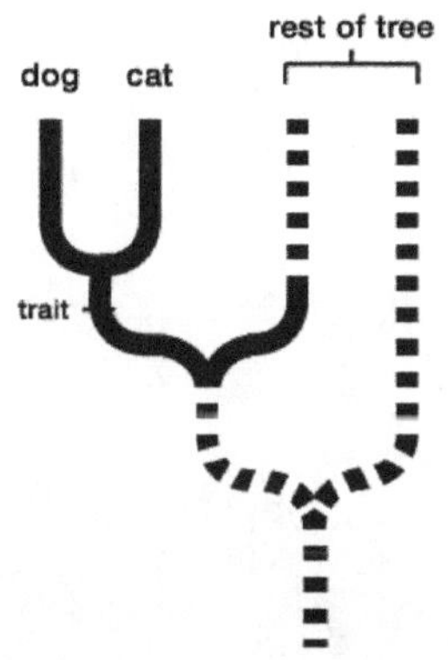

10. What is an amniote, and which animals on the tree in Figure 5.3 are amniotes?

11. If you were to add a cat onto this tree, it would be placed so that the cat and dog are more closely related to each other than to anything else in the tree, as shown. What biological trait could you use in the spot that is marked "trait?"

Puzzle 5.C. Vegetarian Edition

12. What makes the seaweed different from all the other plants on this tree? *Check the box next to the correct answer.*
    - ❑ It has leaves.
    - ❑ It's a vegetable.
    - ❑ It uses spores to reproduce.
    - ❑ It is a photosynthetic autotroph.

13. Consider whether a banana is more closely related to a lemon or an onion. Why might the correct answer be surprising to many people?

Name: ______________________ Lab Time: ______________ Due: ______________

5 Oregon Coast Phylogenetics

This lab will use a collection of "beach gems" to help you gain an appreciation for how biologists approach studying and analyzing organism characteristics to establish evolutionary relationships. Everything you observe in this lab was collected on a beach or in a tide pool in Oregon.

Each group in your lab will work with an identical mixture of organisms. Each individual in the mixture represents a **different species.** Using this group of organisms, you will construct a **phylogeny** (evolutionary history) based on shared ancestral traits. The phylogeny you construct will be a branching tree diagram that represents the evolutionary history (and thus evolutionary relatedness) of this group of organisms. Learn more about phylogenetic trees in chapter 26 of your textbook.

Throughout this lab, it is essential to remember that *phylogenetic trees are hypotheses*. Scientists use the scientific method to create these trees and then use a variety of experiments to test their trees.

Learning Outcomes

- ❑ Explain how an organism's morphology can be used to determine evolutionary relatedness.
- ❑ Create a phylogenetic tree using morphological observations.
- ❑ Utilize the terminology commonly associated with phylogenetic trees.

Activity 5.1. Assembling a Data Matrix

The scientific method starts with observation. In this case, you have a set of organisms represented by their remains and are asking the question, "How are these organisms related?" To begin, you must carefully observe each organism and note similarities and differences in their morphologies. The traits you choose to work with should have an underlying genetic basis. When scientists hypothesize about the evolutionary relatedness of organisms, they often begin by making observations and compiling these observations into a data matrix (Table 5.1). Note that instead of providing a wordy description of each organism, the scientist has chosen straightforward traits and indicated presence or absence with a simple plus "+"/ minus "-" system.

Table 5.1. Data Matrix Comparing Representative Chordates

	Vertebrae	Jaws	Tetrapod (4 limbs)	Hair/Fur	Notochord
Lancelets	–	–	–	–	+
Mammals	+	+	+	+	+
Lizards	+	+	+	–	+
Lampreys	+	–	–	–	+
Bony Fish	+	+	–	–	+

5

Procedure A. Creating a Group Data Matrix

Many of these organisms are fragile and should be handled carefully.

Instructions

1. Carefully spread out your group's specimens on your workspace.
2. Note that each specimen is labeled with a letter. These letters are consistent throughout the class and can be used for reference.
3. **Working as a group**, observe each specimen with regard to its size, shape, texture, and other characteristics.
    a. Color is probably not a good trait to observe, as many of these organisms are not the same color in life as they are when washed up on the beach.
4. Consider using a T-chart to record your observations in your lab notebook. A T-chart is a diagram used to display two different points of view on a topic. In this case, you will divide a page in your lab notebook in half. Label one side "Similarities between organisms" and the other side "Differences between organisms." Refer to specific organisms by letter in the chart.
5. **Working as a group**, decide which features will be most useful in the construction of an evolutionary tree.
6. Use these features to create a group data matrix in Table 5.2 (p. 63). This table should be part of your lab submission.

Before you get off track!

- Consider all of the organisms in your collection to be mature adults.
- Most of these organisms have been sitting in the sun. Their natural color may not be visible.
- Individuals can vary greatly in appearance. Examining different examples of the same organism from various collections can help you distinguish between individual variation and group characteristics.
- Focus on traits that are more likely characteristic of the species rather than of individuals.

Table 5.2. Data Matrix Comparing the Morphology of Oregon Coast Organisms

Organism	A	B	C	D	E	F	G	H	I
Trait ____________									

Name: ______________________ Lab Time: ______________ Due: ______________

5 Applying What You've Learned

A selection of these questions will appear on your lab quiz. Lab quizzes are open-book. Type up answers to these questions in advance so that you can copy and paste the answers into the quiz.

1. Consider all the phylogenetic trees drawn in your class. Remember that the shells used for each group were the same. Are the trees all the same? Explain why or why not.

2. Compare your tree to the DNA-based tree provided by your instructor. Does your tree agree or disagree with this tree? Why or why not?

3. Did your group consider any of the shells provided to be an out-group? If so, which one and why? If not, why?

4. According to your tree, which organisms are most distantly related? Explain your answer.

5. According to your tree, which organisms are most closely related? Explain your answer.

6. Describe one evolutionary trend in invertebrates that is represented by your tree (remember, you can only see the remains, so you can only talk about the remains!). Give specific examples.

7. Explain and describe the types of evolutionary selective forces that may have caused this trend. Be specific and refer to particular shells). These factors could include predation, habitat, environmental conditions, food habits, and others.

8. The collection of organisms provided for this activity contains two species that fall under the common name "limpets." What does the DNA-based tree say about the evolution of limpets? Did your tree agree or disagree with the DNA-based tree with regards to limpets?

Name: ______________________ Lab Time: ______________ Due: ______________

6 Hardy-Weinberg & Natural Selection Pre-Lab

1. Differentiate between the terms evolution and natural selection.

2. Review the procedure for our simulation of Hardy-Weinberg equilibrium. Complete Table 6.1 to explain how each evolutionary process is avoided in the simulation.

Table 6.1. Evolutionary Process in a Simulation

Evolutionary Agent	How is this process overcome in the state of Hardy-Weinberg equilibrium?
Mutation	
Gene Flow	
Genetic Drift	
Natural Selection	
Non-random Mating	

3. Which of the following statements regarding the Hardy-Weinberg equation are true? **Circle the letters for all that apply.**
   a. If $p^2 + 2pq + q^2 = 1$, evolution is taking place.
   b. If $p^2 + 2pq + q^2 = 1$, evolution is not occurring.
   c. If the values of p^2, $2pq$, and q^2 stay the same from one generation to the next, evolution is not occurring.
   d. If the values of p and q change from one generation to the next, evolution is taking place.

4. Complete each of the graphs in Figure 6.1 to show the distribution of alleles most commonly associated with directional, stabilizing, and disruptive selection.

Figure 6.1. Graph the Distribution of Alleles Associated with Selection

Directional	Stabilizing	Disruptive

5. The curves you drew in the previous answer are each the result of a different type of natural selection (remember that natural selection, like all forms of evolution, focuses on change over time). Draw a frequency distribution graph in Figure 6.2 for this population at the start of an experiment to determine what type of natural selection is happening. Be sure to label the axes.

Figure 6.2. Draw a Frequency Distribution Graph to Chart Natural Selection

Name: ____________________ Lab Time: ______________ Due: ______________

6 Hardy-Weinberg & Natural Selection

This lab will focus on the processes that contribute to microevolution by using the Hardy-Weinberg equation to model populations that are not undergoing evolution, and then examine how natural selection occurs.

Learning Outcomes

- ❑ Describe the five processes that contribute to evolution at the population level.
- ❑ Use the Hardy-Weinberg equation to determine allele frequency, genotype frequency, and to determine if evolution is taking place in a population.
- ❑ Differentiate between the three types of natural selection.
- ❑ Describe the role of the environment in evolution.

Activity 6.1. Using the Hardy-Weinberg Equation

Hardy and Weinberg proposed that, in an ideal situation where no evolutionary forces are affecting a population, the frequency of p and q (and those of p^2, q^2, and 2pq) should stay the same from one generation to the next (the **Hardy-Weinberg Principle**). A population that meets these requirements is said to be at **equilibrium**. If, however, frequencies change from one generation to the next, the population is said to be evolving. The mathematical representation of this principle is:

$$p^2 + 2pq + q^2 = 1$$

This activity will use beans to represent individuals that carry two copies of the bean color gene. The bean color gene has two versions (A and a). These alleles interact to produce 3 different phenotypes (red, white and speckled).

- ❑ Red beans represent individuals with two copies of allele A (AA).
- ❑ White beans represent individuals with two copies of allele a (aa).
- ❑ Speckled beans represent individuals with one copy of A and one copy of a (Aa).

Table 6.4. Simulation Data

		Expected # of Offspring *(based on Mendel's Laws)*			Actual Number of Matings Observed *(pulled from the container)*		Number of Offspring Produced *(# expected × total number of crosses)*		
Possible crosses	**Genotypes of the cross**	**# AA offspring expected per cross**	**# Aa offspring expected per cross**	**# aa offspring expected per cross**	**Number of times drawn (tally marks)**	**Total number of crosses**	**# AA offspring produced**	**# Aa offspring produced**	**# aa offspring produced**
Red × Red	AA × AA	4							
Red × Speckled	AA × Aa	2	2						
Red × White	AA × aa		4						
Speckled × Speckled	Aa × Aa	1	2	1					
Speckled × White	Aa × aa		2	2					
White × White	aa × aa			4					
						# Produced by Genotype (100 total)			
						Genotype Frequency			

6

Q10. Did your population evolve, or was your population at equilibrium? How do you know?

Q11. If your population evolved, why? This simulation was designed to represent a population at equilibrium. What went wrong?

6

Procedure B. Simulating Evolution

In this activity, your group will design its own experiment to cause evolution in the bean population.

Instructions

1. Design a simulation to model the impact of different types of evolution on our bean population. Choose one.
   - ❑ Mutation
   - ❑ Gene flow
   - ❑ Genetic drift
   - ❑ Natural selection
   - ❑ Non-random mating

2. Write your question, hypothesis, and prediction on the whiteboard. Return to Part A for a model of how to formulate questions, hypotheses, and predictions for this type of experiment.
3. Add the step-by-step methods for your experiment on the whiteboard.
4. After receiving your instructor's approval, complete Table 6.5.
5. Create a table similar to Table 6.5 in the space provided for Table 6.6. *Revisit the results section of Appendix A for information on the proper formatting of scientific tables.*
6. Run your proposed simulation and record your data in Table 6.6.

Important: Keep a close eye on the beans you add or remove from your population. Make sure that you return your population to its original state before leaving the lab!

Table 6.5. Procedure 6.1.B Experimental Design

Group Question	
Your Hypothesis	
Step-By-Step Procedure	**Complete on Whiteboard**
Prediction	

Table 6.6. Title:

Q12. Did your population evolve, or was your population at equilibrium? How do you know?

Q13. Did your experimental design simulate the evolutionary force you expected? Why or why not?

6

Activity 6.2. Phenotypic Variation & Natural Selection in Mollusks

Many processes can cause the frequency of an allele to change, but the primary cause is **natural selection**. Genetic variability in a population is often assessed with phenotypic frequency histograms.

In an ideal situation, a frequency histogram of the variation of any particular trait would resemble a bell curve, with intermediate phenotypes being most common and extremes being rare.

Changes in the shape of this curve indicate that natural selection is taking place in one of three ways.

- ❑ **Directional selection**: One of the extreme phenotypes increases in frequency while the intermediate phenotypes decrease in frequency.
- ❑ **Disruptive selection**: The intermediate phenotypes become least common, while the extreme phenotypes become more common.
- ❑ **Stabilizing selection**: Intermediate phenotypes increase in frequency while extreme phenotypes decrease in frequency.

Procedure A. Measuring Phenotypes in Mollusks

Instructions

1. Obtain a sample of mollusk shells.
    a. Each sample represents a population of many individuals of the *same species*.
    b. Assume that all the individuals in the sample are the *same age*.
2. Select phenotypic characteristics that can be measured on each specimen. The traits you choose should be:
    a. genetically inherited
    b. related to survival and reproductive success (i.e., operated upon by natural selection)
    c. quantifiable (able to measure with numbers, so color will not work on this one)
3. As a group, decide which trait you would like to focus on.

Confirm your choice with your instructor before proceeding.

6

4. Formulate a hypothesis about what type of natural selection you expect this trait to exhibit (directional, stabilizing, or disruptive) and explain your reasoning.
    a. There will be a fair amount of speculation here, as you are not familiar with these organisms in their natural habitat.
5. Record your hypothesis and prediction in your lab notebook.
6. Divide the shells into 7-10 categories based on the trait chosen by your group.
    a. Numerically define the range of each category.
    b. Record the number of individuals in each category.
7. Define the range of your phenotype (if you are measuring size, the smallest is one end of the range and the largest is the other end). Divide this range into an odd number of subgroups (try to stick to multiples of 5 or 10; e.g., 0–5 mm is subgroup 1). Aim for 7–11 subgroups within your range.
8. Create a table in your lab notebook to record your data. In the cells of the table, you will record the number of individuals in each subgroup with tally marks.
9. Use available materials (digital calipers, rulers, balance, etc.) to measure that characteristic for each specimen in your sample. Record your data in your table.
10. Determine the median value for the characteristic you have chosen.
    a. List your subgroups from biggest to smallest.
    b. The subgroup that falls in the middle is your median.
11. Create a graph (Figure 6.3, p. 77) with the range of character states (subgroups) on the X-axis and the number of individuals within a subgroup on the Y-axis.
    a. This graph will ultimately have a line (drawn relative to the Y-axis) representing the "before" distribution of the character you are studying, bars corresponding to the actual frequency distribution of your data, and a second line based on the frequency distribution showing the "after" distribution of the character you are studying.
12. Draw a line near the top of your graph showing the relative "before" distribution of that trait in your population.
    a. To do this, we will assume that the median subgroup of your range was originally the most common allele. Your line should show a bell curve peaking in that middle subgroup.

13. Add bars to the graph to create a frequency distribution for the trait you measured.
    a. A frequency distribution shows the number of individuals exhibiting each variation (or range) of a characteristic.
    b. Do not graph individual data points (our sample size is too small for this to work).
    c. Each bar on your graph will represent one subgroup within your defined range.
14. Draw a curve on the graph that represents a "best fit" of your frequency distribution. This is your "after" distribution.

Figure 6.3. Before, Actual, and After Data Distribution of a Mollusk

6

Name: ______________________ Lab Time: ______________ Due: ______________

7 Bacterial Diversity Pre-Lab

Aseptic Technique

1. Why is it important to use aseptic techniques when working with bacteria?

2. List three ways you can practice aseptic technique in the lab.

3. What is an inoculation loop? ____

Gram Stain

4. What's the difference between the plasma membrane and a cell wall?

5. In the space below, sketch the cell wall of both Gram-positive and Gram-negative bacteria. Your sketch need not be extremely detailed, but should highlight the differences between the two walls enough that you could use them to answer questions about wall structure.

6. Indicate the order in which you use the following reagents in the Gram stain process (put the number one next to the reagent you use first, two next to the second, etc.).

    ___________ 95% ethanol

    ___________ Crystal violet

    ___________ Gram's Iodine

    ___________ Safranin

Selective and Differential Assays

7. Explain how an agar plate can be both selective and differential.

8. Differentiate between TSA, MAC, and MSA agar in terms of their suitability for growing bacteria.

9. Which of these methods can confirm the results of a Gram stain? ___________________

10. What does it mean if a species does not grow on an MSA plate?

11. If a species does not grow on an MSA plate, can it provide insights into that species' Mannitol metabolism?

Name: ____________________ Lab Time: ______________ Due: ______________

7 Bacterial Diversity

Prokaryotic cells are more prolific and widespread than any other taxonomic category in the tree of life. However, we have limited knowledge of the actual number of species that exist. Prokaryotes are tiny (usually 1–2 μm) and can be difficult to culture in a laboratory. More common bacteria, such as those found on the surfaces of our bodies and in our environment, are easier to study.

In general, scientists rely on morphological and biochemical traits (and DNA) to identify these organisms. In this lab, you will learn to identify bacteria by the shape of their cells, the type of colony they form, the structure of their cell wall, and some of their nutritional habits.

Learning Outcomes

- ❑ Safely and aseptically handle bacterial cultures.
- ❑ Describe several ways that bacterial species can be categorized.
- ❑ Relate the results of Gram stain testing to the structure of the bacterial cell wall.
- ❑ Create a phylogenetic tree using observations of several bacterial species.

Activity 7.1. Bacterial Morphology

One way to classify bacteria is based on their morphology, specifically their shape. Most bacteria exhibit one of three shapes: bacilli (hot-dog), cocci (spheres), and spirilla (corkscrews) (Figure 7.1). Bacteria can also be found individually in an environment or organized into an aggregate or chains. The individual cells will not look much bigger than a grain of sand. If you do find a chain of cells, look closely to see the individual cells within that chain.

Figure 7.1. The Three Shapes of Bacterial Cells. *Made with Biorender.*

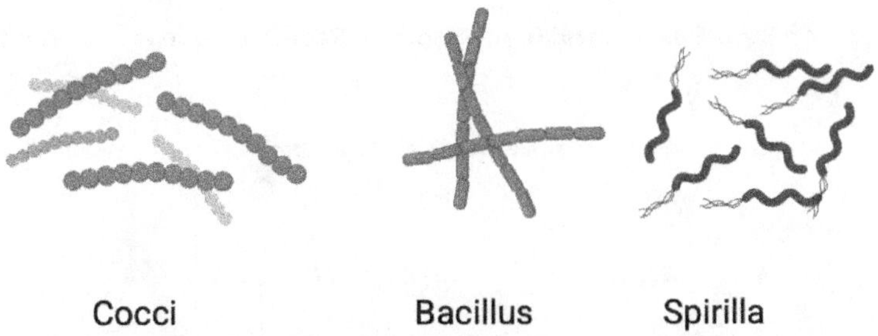

Hints

- Bacteria are very small! Start your search by focusing on the edge of the coverslip and then move around the slide at low magnification until you see a colored specimen (pink or purple). Only after finding a stained region should you increase the magnification.
- Keep the size of bacteria in mind. If you've found something that is easy to see, it's probably just a bit of debris.
- Contrast is critical when viewing prokaryotes. Use the iris diaphragm to adjust contrast.
- To visualize staining in bacterial cells, use a slightly lower contrast and a bit more light than when examining unstained cells.

Procedure A. Identifying Bacterial Shapes

Instructions

1. Obtain a "Bacterial Types" slide. Some of these slides (round coverslips) have a mixture of all three shapes in the same place on the slide, while others (rectangular coverslips) have three different drops that you must navigate between to see all the shapes.
2. Use the hints above to focus on the bacteria with your microscope. Remember to start with the 4× objective and work your way up to the 40× objective rather than just starting at high power.
3. Find examples of the three bacterial shapes.
4. Create a scaled sketch of each of the shapes in one section of your lab notebook. Remember to include a scale bar.

Activity 7.2. Learning the Gram Stain Technique

Another way to categorize bacteria is by their cell wall structure. The primary building block of the bacterial wall is peptidoglycan, a complex polymer. The Gram stain is based on the fact that a particular dye, crystal violet, can be permanently attached to exposed peptidoglycan.

Differences in structure between the cell walls of Gram-positive and Gram-negative bacteria (Figure 7.2) affect the staining results. After Gram staining, the cells of Gram-positive bacteria are purple and those of Gram-negative bacteria are pink.

Figure 7.2. Comparison of the Cell Wall Structure of Gram-Positive and Gram-Negative Bacteria. *Made with Biorender.*

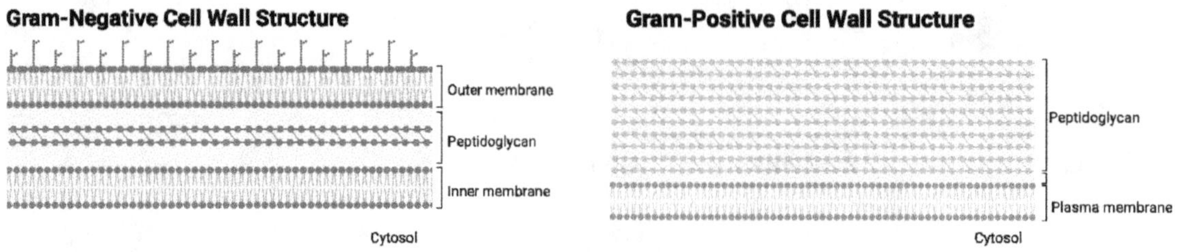

Practicing the Gram Stain

Gram staining requires four different chemicals, each with a different purpose.

- ❑ **Crystal violet**: Attaches to all cells (purple).
- ❑ **Gram's Iodine**: Creates a chemical bond between crystal violet and the peptidoglycan.
- ❑ **Alcohol**: Removes excess (and unattached) crystal violet.
- ❑ **Safranin**: Attaches to all cells (pink).

In this activity, you will practice performing the Gram stain by preparing smears of your gum tissue.

Instructions

1. Place a single, tiny drop of water in the center of the slide.
    a. If you get too much water, blot the edge of the slide with a tissue. Too much water results in a longer drying time.
2. Use a permanent marker to draw a capital "E" on the side of the slide to use as a reference point.
    a. It is easy to accidentally flip over the slide in this process, and the letter "E" will help you orient the right side up.
3. **Gently** rub your gingiva (gums) with a toothpick.
4. Smear the end of your toothpick into the water drop and along the ENTIRE slide to spread out and create a thin smear. Use the toothpick like a squeegee to make a very thin smear (film).
5. Air-dry the smear.
6. Attach a clothespin to your slide.
7. Set up a Gram stain station at your bench on your lab mat. The staining tray should be in the middle of your mat. "Contain the stain!" by keeping the staining bottles on the lab mat or in the kit bucket.
    a. If you do spill a stain on the countertop or floor, please use Ethanol to clean it up.
8. Perform a Gram stain on your gum smear slide. (Future reference: Quick Instructions on p. 84):
    a. Cover your entire smear with CRYSTAL VIOLET for 60 seconds.
    b. Gently wash with water for ~2 seconds.
    c. Cover your entire smear with GRAM'S IODINE for 1 minute.
    d. Gently wash with water for ~2 seconds.
    e. Decolorize with 95% ETHANOL for a MAXIMUM of 20 seconds.
    f. QUICKLY wash the slide with water for ~5 seconds.
    g. Cover your smear with SAFRANIN for 1 minute.
    h. Gently wash with water for ~2 seconds.
9. Remove the clothespin from the slide.
10. Blot your slide with Bibulous paper using the following instructions:
    a. Lay the booklet of BIBULOUS PAPER on the countertop. (Holding the book while blotting the slide will result in a broken slide.)

Colony Morphology

Bacteria multiply rapidly. In the laboratory, dilute cultures of bacteria are applied to petri dishes filled with agar (a Jell-O-like substance derived from kelp) mixed with some kind of food source. These petri dishes are called plates. Microbiologists use a variety of different plates to culture prokaryotes (Figure 7.3).

Figure 7.3. A Plate Culturing A Prokaryote Colony

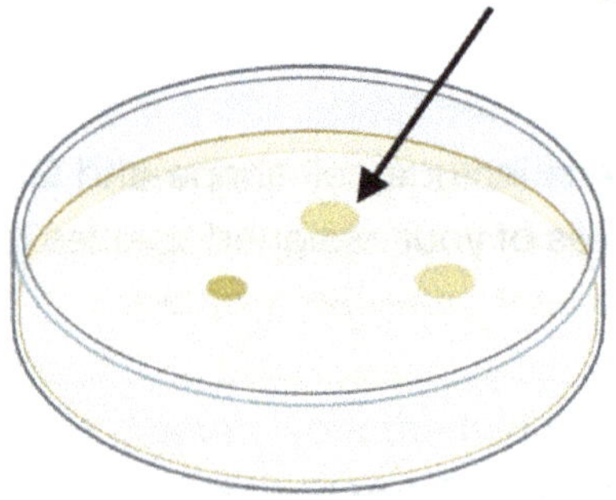

Tryptic soy agar (TSA) is a versatile growth medium that provides all the necessary nutrients for many bacteria to survive and reproduce. Once applied to a TSA plate, a single bacterium can become thousands overnight! On the plate, the offspring produced by the binary fission of a single bacterium forms a circular region of growth known as a colony. Colonies exhibit diverse morphological features, including color, form (shape), margins (edges), elevation (height), and texture, which vary depending on the species that forms them.

Instructions

1. Examine the TSA plate provided for your species using both your eyes and a dissecting microscope.
2. Use the provided handout to characterize the form, elevation, and margins of the colony.
3. Note the color of the colonies and any texture (smooth, rough, shiny, dull).
4. Record your observations in the colony morphology row of Table 7.1.

MacConkey Agar

Prokaryotes can also be classified by their nutritional needs. MacConkey agar (MAC) is a selective and differential growth medium often used in bacterial identification. **Selective** refers to the fact that only certain species can grow on this medium. MAC plates contain bile salts (a component of the bile produced by the gallbladder) and crystal violet, two substances that inhibit the growth of gram-positive bacteria.

Differential refers to the fact that different species will grow differently on this plate. MAC plates also contain lactose, a carbohydrate that certain species of bacteria can use for energy, and a pH indicator that is pink under acidic conditions. Bacteria that use lactose release an acidic product and, therefore, will produce pink colonies on this type of plate. Bacteria that grow on MAC plates but cannot utilize lactose will typically form normally colored or colorless colonies. So, this type of plate "differentiates" between species that can and cannot use lactose as an energy source.

Instructions

1. Examine the MacConkey agar plate provided for your species.
2. Record your observations in the MacConkey Agar row of Table 7.1.

Mannitol Salt Agar

Mannitol salt agar (MSA) is also a selective and differential growth medium. This growth medium contains a high concentration of salt (7.5%), mannitol (a carbohydrate), and a pH indicator. When the bacteria growing on these plates use mannitol, they will produce an acidic product. In this case, however, the growth medium (not the colonies) will change color when a species that utilizes mannitol is present.

Instructions

1. Examine the Mannitol Salt plate provided for your species.
2. Record your observations in the Mannitol Salt Agar row of Table 7.1.

Q3. Bacteria come in only three shapes, and colonies can only have certain forms, but prokaryotes exhibit a variety of metabolic differences. Should cell shape and colony characteristics be a primary determinant in building your phylogenetic tree?

Q4. Based on this information, what type of bacteria does MAC "select" for?

Q5. Explain what is being selected for and what is being differentiated when bacteria are cultured on mannitol salt agar.

Clean Up

When you have finished characterizing your assigned species, clean up!

Instructions

1. Place all plates in the designated biohazard waste container.
2. Return liquid cultures to the class rack for sterilization.
3. Sterilize or dispose of (biohazard waste) any tools that have touched bacterial cultures.
4. Fold up your disposable lab mat and place it in the designated biohazard waste container.
5. Return your supplies to the group kit.
6. Wipe down the body and stage of your microscope and any other equipment used with a towel saturated with disinfectant (do not use this on the lenses of the microscopes, though!).
7. Saturate your work area with disinfectant spray and allow it to air dry or sit for at least 5 minutes.
8. Wash your hands with soap and water.

7

Activity 7.4. Creating a Phylogenetic Tree

We will use the data gathered in Activity 7.3 to build a phylogenetic tree for the six species characterized by the class. You may need to build this tree outside of class, but make sure you gather the class data before leaving today.

Procedure A. Building an Evolutionary Tree

1. Add your observations in Table 7.1 to the class data table (Table 7.2).
2. Individually or as a group, build an evolutionary tree of the bacterial species examined in this lab.
3. Label the shared derived traits (obtained from Table 7.2) unique to each branch you create.

> **HINT**: Think about what you know about bacteria and review the location of bacterial clades on the phylogenetic tree of life. What characteristics are more likely to be part of their genome (and subject to evolution) versus determined by the environment (not subject to evolution)? Choose key heritable traits as the basis for your hypothesis.

Table 7.1. Characteristics of ________________________ (insert your assigned species here!)

Characteristic	Description			Color	
Cell Shape			Gram Stain		
Colony Morphology (TSA plate)	Color	Form	Margins	Elevation	Texture
MacConkey Agar	Growth (+/-)			Colony Color	
Mannitol Salt Agar	Growth (+/-)			Medium color after growth	

Table 7.2. Characteristics of Bacterial Species Examined in BI223z

Species	Cell Shape	Cell Wall Structure (Gram +/-)	Colony Morphology					Lactose Metabolism (+/-)	Salt Tolerance (+/-)	Mannitol Metabolism (+/-)
			Color	Form	Margin	Elevation	Texture			

7

Name: ______________________ Lab Time: ______________ Due: ________________

7 Applying What You've Learned

A selection of these questions will appear on your lab quiz. Lab quizzes are open-book. Type up answers to these questions in advance so that you can copy and paste the answers into the quiz.

1. Explain why the gram-positive wall stains with crystal violet, but the gram-negative wall does not.

2. Would a gingival cell ever stain Gram-positive? Why or why not?

3. In this lab, we determine the Gram stain by taking the "best 2 out of 3" results. Is there a better way to determine if you are getting accurate results? Explain.

4. Which bacterial species are most closely related in your phylogenetic tree? Why do you feel these species are closely related? What traits do they share? What traits make these species different from other species? Refer specifically to the tree you (or your group) created.

5. Which characters were most important in helping you determine the relatedness of the different bacterial species as you formed your tree? Were there features about these characters that make them more important? For instance, do these features play an important role in how the organism functions?

6. Which characters were least helpful? Explain why. Were there features about these characters that make them less important? For instance, are these features unrelated to survival?

7. Examine a diagram of the bacteria branch of the tree of life (like the one we used in this week's Scientific Explanations). We did not perform the Gram stain on cyanobacteria, but can you predict whether cyanobacteria might be gram-positive or gram-negative based on the evolutionary tree? Explain your answer.

7

Name: ______________________ Lab Time: ______________ Due: ______________

8 Invertebrate Communities Pre-Lab

To prepare for this lab, watch the *Oregon Field Guide* episode "Tiny Bugs" (start at 17:40) and answer the following questions. Access the video at http://go.chemeketa.edu/oregonfieldguide or by scanning the QR code.

1. Do some research on Burlese Funnels. How do these experimental apparatuses function?

2. Describe the ecological roles and provide examples of each of the following types of organisms.

Table 8.1. Organisms and Their Ecological Roles

Organism Type	Ecological Role	Example
Carnivore		
Decomposer		
Detritivore		
Fungivore		
Herbivore		

3. Provide at least three biotic or abiotic factors that impact the diversity of a leaf litter community. For each, provide a 3–5 sentence explanation of why you feel this factor will impact diversity.

4. Based on your answer to the previous question, which of the communities in Table 8.2 would you predict to have the highest diversity of invertebrates? Explain your reasoning.

8

Table 8.2. Predicting the Highest Diversity of Invertebrate Communities

A	A forest dominated by a single species of deciduous tree, sparse ground vegetation, and litter consisting of leaves and twigs.
B	A mixed forest dominated by Douglas fir, which is wet for most of the year, features litter containing fir needles, decomposing wood, leaves, and mosses.
C	A wet, grassy field littered with a thin layer of decomposing grass and a few leaves from a nearby forest.
D	A dry, grassy field most of the year, grazed by animals in the recent past and subjected to a controlled burn 6 years ago, the litter consists of a thin layer of grass.

Name: ______________________________ Lab Time: __________________ Due: ____________________

8 Invertebrate Communities

In this lab, we will examine the diversity of invertebrate communities in samples collected at various locations using a Berlese (Tullgren) funnel. The basic setup is to put a soil/litter sample in a funnel placed over a jar containing alcohol (a preservative). A bright light is placed over the funnel. Most soil and litter invertebrates have a negative reaction to light, heat, and desiccation, and will move away from the light, eventually falling into the jar of alcohol. Using our knowledge of the habitats where each sample was collected, we can quantitatively evaluate the invertebrate communities from different habitats (the same size sample was taken from each habitat).

Learning Outcomes

- ❑ Describe the macro- and microscopic diversity of invertebrates in leaf litter.
- ❑ Describe a community in terms of species diversity (relative abundance and species richness).
- ❑ Explain how abiotic and biotic factors contribute to diversity in a given community.

Activity 8.1. Sample Preparation & Predictions

Each group will be provided with a litter sample and an alcohol solution containing organisms gathered from that litter with a Berlese funnel. Samples were collected from about 1 m^2 plots, and organisms were collected with the aid of Tullgren/Berlese funnels. The first step is to concentrate the organisms in the alcohol onto a piece of filter paper for analysis and identification.

Your group will be assigned one of the samples to analyze.

Instructions

1. Place two layers of coffee filters in a cone filter on top of an Erlenmeyer flask.
2. Gradually pour the filtrate from your leaf litter sample over the filter.
3. Use the squirt bottles to spray the inside of the flask, ensuring you collect all the organisms.
4. The filtration process may take considerable time. Be patient!
5. Place the filter paper on a large watch glass and allow it to dry briefly.

3. After exploring and sorting all the organisms on your filter paper, count up the number of individuals of each type found in your sample and record your data in Table 8.4. Tally the total number of individuals of each type found in your sample.
4. Identify the number of morphologically distinct species in each clade. Place that number in parentheses after your total in Table 8.4.
5. Add your data to the class spreadsheet to calculate the Simpson's Diversity Index for your site. The link to that spreadsheet is located on the lab assignment page in Canvas.

Table 8.4. Organisms Found in Leaf Litter from ____________________________ (site name)

Common Name of Order	# Individuals	Total individuals counted	Number of Types (Species) Represented
Ants			
Aphid			
Beetles			
Caterpillars			
Centipedes			
Earthworms			
Flies			
Harvestmen			
Millipedes			
Mites			
Pill bugs			
Proturans			
Pseudo-scorpion			
Roundworms			
Slugs			
Snails			
Spiders			
Springtails			
True Bugs			

8

Q3. Which part of the data recorded in Table 8.4 represents species richness, and which represents relative abundance?

Q4. In the next activity, you will use Simpson's Diversity index to compare the diversity of each site. Why can't we just use species richness?

Activity 8.4. Data Analysis

Use data from the entire class to create the following graphics.

1. Table displaying species richness and Simpson's Diversity Index for each site.
   a. Remember that the numbers in parentheses account for the number of morphological species found in each clade. These numbers should be taken into account when making your calculations.
2. Stacked bar graph comparing the relative abundance of the three most common clades overall at each site.
   a. Determine the three most common clades overall by looking at the total number of individuals across all sites.
   b. Site name on the x-axis; % of total number of species (divide the number of individuals of that species by the total number of individuals) at the site on the y-axis.
   c. The columns should contain data for the same three clades at all sites.
   d. Since you are not using all your data, your totals may not equal 100%.

For information on proper formatting for graphs and tables, consult Appendix A.

Q5. Does the data gathered during this lab support your hypothesis (pre-lab) and prediction (Activity 8.1)? Include a scientific explanation.

Using Published Material to Substantiate Facts

When to Use

- Background section of the introduction
- When stating a fact or explaining a concept that is not covered in your textbook
- Whenever a specific number or date is used in the text
- When discussing species-specific information
- When new information is brought into the discussion

How to Incorporate

- In-text citation (see references section of this document)

Using Published Material to Support Arguments and Interpretations

When to Use

- In the introduction, when explaining how you came up with your hypothesis
- In the discussion, when explaining what you think your data means

How to Incorporate

- Write a summary of the study performed. What was the experiment, what were the results, and how are those results relevant to your work?
- Include an in-text citation at the end of your description.
- If the description is long, you may need to include a citation at the beginning and end of your summary.

A

Title

The title of a paper seems like an insignificant thing. Who cares about the title as long as the rest of the paper is good, right? Think again! Titles can be one of the most important parts of a scientific paper. Generally, scientists publish their papers in scientific journals where papers are listed by title in the table of contents. Your goal as a scientist should be to create an informative but interesting title that will encourage the readers of the journal to turn to the first page of your article.

Writing a Title

In terms of content, the title of a scientific paper should summarize in one sentence or less the overall theme of the paper, including the independent and dependent variables. Indicate the question being answered and hint at the conclusion. Beware of titles that seem too short or too long!

Short Titles are Usually Too General (and lack one or both variables)

Fava Bean Growth

What aspect of fava bean growth interests you? Why is fava bean growth interesting? Why would I want to learn more about it?

Long Titles are Generally Repetitive

Organic Fertilizers including Manure and Chemical Fertilizers like Rapid Grow Cause Fava Beans to Grow More Rapidly than Untreated Fava Beans but Chemical Fertilizers Do Not Cause Fava Beans to Grow Faster than Organic Fertilizers

Did you make it through that one? It's challenging to capture and keep a reader's attention with such a long title. Not only does this title provide too much information, but it is also very repetitive.

Try This Title

Organic and Chemical Fertilizers are Equally Efficient at Increasing the Growth of Fava Bean (Vicia faba) Seedlings

Long enough to tell you what's going on and imply the question the paper is focusing on, but short enough that a reader can get through it. That's your goal.

A

General Suggestions for the Title

- Only the first letter of the title and proper nouns should be capitalized.
- Usually, scientific names (not common names) are listed in the title. Remember, a scientific name is always underlined or italicized. Genus is always capitalized, and species is not.
- The title should be a slightly larger font size than the rest of the text and can be highlighted by using bold type.

Introduction

This section of the paper provides the reader with an Introduction to the topic being discussed and focuses the reader on specific aspects of this topic. The introduction ends with a statement of your hypothesis and a prediction of what your results will be.

Introductions begin with general information and become very specific toward the end. The text should begin with basic background information and work slowly toward the specifics of the

Results

The Results section is where you present your data to the reader in a completely objective manner. Separating the data and your interpretation of the data allows readers to draw their own conclusions, but you don't want your readers to draw just any conclusion. You want them to conclude the same thing that you concluded. To accomplish this, your writing must act as a guide for the reader, highlighting key pieces of data and effectively leading the reader to the conclusion you want them to draw. The Results section of a paper gives the reader **only the facts** of the experimental results. The facts are presented in three ways: figures, tables, and text.

Tables & Figures

A scientific paper presents data in two formats: tables and figures. A table is a chart based on rows and columns where data, numeric or otherwise, is organized by category. A figure is any type of visual data display that does not involve a table. Figures include graphs, photographs, diagrams, and drawings. Figures and tables are numbered separately in a formal scientific paper.

Tables vs. Graphs

Some of the types of figures are obvious, but you may find the use of tables and graphs confusing. We often collect data in the form of a table because it is easy to keep the data points organized; however, this is not usually the best way to visualize how the dependent variable is affected by a change in the independent variable.

A

> Insert box for emphasis: Figures and Tables should be labeled separately and consecutively (Figure 1, 2, etc and Table 1, 2, etc.)

Use a Table When:

- Neither your dependent nor your independent variable is quantitative.
- You are displaying single data points for each treatment.
- Your overall number of data points is small (5 or fewer).
- When the data points are not continuous.

Use a Graph When:

- At least one of your variables is numeric.
- Your data points are continuous.
- You are displaying more than one independent variable.
- You want to show a trend or rate of change.
- You have many data points (more than 5).

Bar Graphs vs. Line Graphs

If you have decided to display your data in a graph, you have one more decision to make. What type of graph should you create? Many types of graphs exist, but most often, you will need to choose between a bar graph and a line graph.

Use a Line Graph When:

- Both variables, independent and dependent, are numeric.
- The data is continuous.
- The differences between data points occur on a small scale (you can adjust the scale on the axis to make the differences more obvious).
- More than one independent variable is being plotted on the same graph.

Use a Bar Graph When:

- One of your variables is not a numeric value.
- The data is non-continuous.
- The data is continuous, and the change between data points is significant.
- Categories need to be displayed in relation to one another.

> Notice that there is some overlap between these graph types. The key is to determine which format best displays the trends you see in the data.

Graph Format Requirements

- The **independent variable** always goes on the X-axis.
- The **dependent variable** always goes on the Y-axis.
- Choose a scale that is appropriate for each axis (your data points should fill your graph rather than being squashed together at one end).
- The X- and Y-axes should be clearly labeled, and units of measurement should be indicated.
- If a graph has more than one line, the different lines should be clearly distinguishable by color and/or line texture, and a key should be provided.
- Numbers should not be attached to data points.

Figure Legend Requirements

All figures and tables should include a **descriptive legend** that includes a brief description (1–2 sentence) of the table or figure. The description should be focused on what the reader needs to know to understand the figure. When writing a figure Legend:

- Figure numbers should be sequential and ordered based on when they are referred to in the text.
- Include a one-sentence title in bold that states the take away message of the figure.

- Use what you know about how life works and how your system works to make sense of the results.
- Lead the reader to your conclusion rather than allowing them to draw their own conclusion.

Begin your discussion by restating your hypothesis and explaining whether your data supports or refutes the hypothesis.

> We proposed that the application of organic fertilizers would increase the growth rate of fava beans more than synthetic fertilizers. Although both types of fertilizer increase the rate of seedling growth compared to unfertilized seedlings, fava bean seedlings treated with synthetic or organic fertilizers grew at approximately the same rate and were approximately the same size after 5 days (figure A.2). This does not agree with previous results that show organic fertilizers produce larger crop yields in tomatoes (Smeagol et al., 1998) and strawberries (Hallo and Schmidt, 1999). Unfertilized seedlings stopped growing after 3 days.

The meat of the discussion really has two parts. First, explain why you think you got the results that you got. Second, use previously published scientific evidence to support that explanation.

- You do not need data that exactly matches yours.
- You do need to demonstrate that your interpretation is scientifically valid.

A

> During germination, storage proteins, lipids, and carbohydrates stored in the cotyledons of the fava bean are used to power and support the growth of the seedling. Once photosynthesis begins, seedlings absorb water and inorganic nutrients from the soil. Nea and Popper (2012) used radioactive nitrogen to track the flow of nutrients into sunflower seedlings. They found that 3 days after photosynthesis began. At the same time, radioactive nitrogen was still present in seedling cotyledons; it ceased moving from the cotyledon into the seedling and began to flow into the seedling from the soil. This explains why seedlings grown in sand without fertilizer stopped growing after 3 days. As the seedlings in this experimental group depleted their stored nutrients, they were unable to absorb additional nutrients from the soil (since it lacked all forms of fertilizer). They could no longer obtain the building blocks necessary for synthesizing new plant tissues.
>
> The similarity in the results of the two fertilizer treatments may be a result of the concentration of fertilizer used. Synthetic fertilizers contain more concentrated nutrients than organic fertilizers (Porter, 2011); however, in this experiment, we used the same amount (by weight) of both types of fertilizers. Determining the actual concentration of key nutrients, such as nitrates and phosphates, in synthetic and organic fertilizers would allow us to apply them at the same concentration to the soil and ensure that differences between treatments were due to the type of fertilizer, rather than the concentration of the fertilizer.

Then, address any limitations in the experiment(s). Limitations are not necessarily errors or flaws. You can only investigate one small piece of the puzzle in any one experiment. What new experiment would expand your ability to understand this process and/or make your results more applicable?

> These results were limited by the fact that only one type of organic fertilizer was tested. Most organic farmers use a combination of three to five different fertilizers to produce maximum crop yield (Cenote, 1999). Different organic fertilizers provide different types of nutrients. Bone meal, for example, is high in phosphorus and has some nitrogen (Cenote, 1999). Future experiments should be designed to examine seedling growth in Fava beans treated with combinations of manure, bone meal, bat guano, and alfalfa meal, ensuring that seedlings receive the optimal combination of nutrients.

Finally, end the discussion section with a concluding paragraph.

> Our results suggest that organic and chemical fertilizers are equally effective in promoting plant growth. However, further research is necessary to determine the optimal concentration of fertilizer required and whether other organic fertilizers or combinations of organic fertilizers will have a similar effect.

Notice that you can use the first person in the discussion section. It is best, however, to limit its usage. The data should stand on its own—you're just interpreting it!

General Suggestions for the Discussion Section

- Always use the past tense.
- Refer (again) to all your figures as you explain them.
- Use peer-reviewed references to support what you say.
- Assume your reader knows nothing. At this level, it is essential for you to explain exactly why you got the results you did (or why your results did not match your hypothesis). For example, if heat increases the rate of a reaction, explain why (at the molecular level).
- Ensure that you interpret all your data accurately. For instance, if there's a flat portion of a graph, tell your reader why it is there or at least provide a hypothesis. Explain whether you expected these results and why or why not.

References

Any information that is not considered **standard knowledge** must be cited and referenced in a scientific paper. As a general rule, consider anything found in your biology textbook as standard knowledge in the field of science. Wherever you go into more depth, you need to provide a reference.

Scientists do not quote directly from papers but summarize the facts or conclusions of a paper in one or a few sentences that are referenced at the end of a sentence. You must practice

General Suggestions for the References Section

- Make sure that the in-text citation matches the start of the associated reference in the reference list. This is how the reader connects the citation to the actual source.
- Italicize all scientific names used in the references section.
- Book and journal titles are always italicized or underlined.
- Chapter titles and article titles are in normal font, and only the first letter of the first word is capitalized.

Peer Review & Publication

Scientists do not work alone; they work as part of a team. Although there is some competition in science, most scientists prefer to work with their colleagues by sharing their ideas and data and receiving feedback. The publication of scientific papers in a respected journal is a crucial step in the scientific process, as it makes your work accessible to the public. Once a paper is published, a larger number of scientists are available to provide additional feedback and ideas to further your research.

Peer review is a process that ensures the validity of your work. After submitting a scientific paper to a journal, the journal will identify experts in the field related to the paper and send it to them for critique. When you read a paper that has been published after peer review, you can be confident that the logic, procedure, and ideas presented have been thoroughly vetted by a group of experts who are familiar with that topic. That being said, reviewers do make mistakes! Occasionally, a paper that has passed through the peer-review process is later retracted.

A

Throughout your biology course, you will be asked to peer-review the work of your teammates and other members of your class. Being able to provide constructive criticism in these reviews (and being able to handle the criticism you receive in return) is all part of the scientific process.

Tips on Providing Constructive Criticism

- **Remember that your goal is to help the writer produce a better product**, rather than making them feel that they are unprepared or unskilled.
- **Provide both praise and critique in your comments.** Point out to the writer what you like about their work before you start addressing your concerns.
- **Be specific in your praise.** Telling the writer that you like their work overall will not mean as much as when you say you liked a specific part of the introduction. Being specific shows that you really read the paper.
- **Be specific in your feedback.** Instead of saying that the introduction doesn't make sense, point to a particular paragraph or section that confuses you and explain why.

- **Ask questions as well as making comments.** Whenever possible, use comments like "Did you mean…" or "Would …be a better word to use here?" Questions are a less confrontational way to get your point across and provide the opportunity for the writer to correct things on their own.
- **Remember that it's okay to admit you don't know how to fix something.** Point out your concern, but acknowledge to the writer when you don't have a quick fix.

Tips on Receiving Constructive Criticism

- **Read all of the feedback provided before reacting**, then sit back and let it sink in for a minute.
- **Don't take the feedback personally.** Remember that your reviewer is trying to help you get a better grade, not hurt you.
- **Remember that you don't have to do everything suggested by your reviewer.** We use more than one reviewer so we can get different opinions. If all your reviewers say the same thing, then you should take the comment seriously, but if only one person commented on a specific item, ask yourself whether the comment is key to your success or if it is a point of personal preference on the part of the reviewer.
- **Start with the small stuff!** Fix things like typos first, and formatting first, then move into the more major changes.

Acknowledgments

Unless otherwise stated, figures created by Jennifer Schramm or Chemeketa Press.

Figure 5.1. "Nova Evolution Lab," produced by NOVA. PBS LearningMedia.

Figure 5.2. "Nova Evolution Lab," produced by NOVA. PBS LearningMedia.

Figure 5.3. "Nova Evolution Lab," produced by NOVA. PBS LearningMedia.

A

www.ingramcontent.com/pod-product-compliance
Lightning Source LLC
Chambersburg PA
CBHW080335300826
50559CB00001B/4